How the Great Scientists Reasoned

How the Great Scientists Reasoned

The Scientific Method in Action

Gary G. Tibbetts
Physics Department
General Motors Research Laboratories
Warren, MI, USA

ELSEVIER

AMSTERDAM • BOSTON • HEIDELBERG • LONDON • NEW YORK • OXFORD
PARIS • SAN DIEGO • SAN FRANCISCO • SINGAPORE • SYDNEY • TOKYO

Elsevier
32 Jamestown Road, London NW1 7BY
225 Wyman Street, Waltham, MA 02451, USA

First edition 2013

British Library Cataloguing-in-Publication Data
A catalogue record for this book is available from the British Library

Library of Congress Cataloging-in-Publication Data
A catalog record for this book is available from the Library of Congress

ISBN: 978-0-323-28267-3

For information on all Elsevier publications visit our website at store.elsevier.com

This book has been manufactured using Print On Demand technology. Each copy is produced to order and is limited to black ink. The online version of this book will show color figures where appropriate.

Dedication

To my wife, Pat

Contents

Acknowledgments

Sincere thanks to those friends and colleagues from General Motors Research Laboratories who have read the chapters of this manuscript: Jack Gay, Rich Teets, Cameron Dasch, Jeff Sell, and Harvey Burley. Their careful proofing and useful suggestions have been vital and have saved me from many errors. I'm especially grateful to Rich Teets for helping me to straighten out the logical threads leading to Planck's radiation law.

Thanks to my friend and filamentous carbon colleague Carlos Bernardo of the University of Minho, Portugal, for giving me useful corrections and the Portuguese slant on Chapter 3 on Columbus. Thanks also to António Lázario, the University of Minho, for suggesting some modern historical interpretations for that chapter.

Much appreciation is due to my daughters Meg, Elizabeth, and Kate and to my sons-in-law Andrew and Bryon for reading manuscript chapters, wordsmithing and providing technical advice. My wife Pat has my sincere gratitude for adding vital focus to this book.

1 Introduction: Humanity's Urge to Understand

In the beginning was the book of Nature. For eon after eon, the pages of the book turned with no human to read them. No eye wondered at the ignition of the sun, the coagulation of the earth, the birth of the moon, the solidification of a terrestrial continent, or the filling of the seas. Yet when the first primitive algae evolved to float on the waters of this ocean, a promise was born—a hope that someday all the richness and variety of the phenomena of the universe would be read with appreciative eyes.

Perhaps the earliest traces of human thought remaining are works of art depicting animals and the hunt and celebrating fertility, but our earliest record of human speculative thought comes to us from stories told and retold around the campfires of the ancient world. These myths often were primarily concerned with the facts of nature. How was the earth created, and what was the starry firmament? Why did winter follow summer? What was the rainbow? These myths testify that deep within our nature is the urge to understand.

In the pages to follow, we will see that the understanding of our physical world begins with something related to the myth: the hypothesis. They share a common ancestor because a hypothesis can be little more sophisticated than a flat-out guess about what causes a phenomenon. In the thousands of years it took human curiosity to evolve from mythical fabrication to experimentally supported conclusions, the hypothesis has evolved to be the primal tool for uncovering the deepest secrets of nature. We will describe how our predecessors evolved helpful and sophisticated modes of thought to test and distinguish valid hypotheses from invalid.

Chapter 2 will present a brief summary of the elements of the scientific method, which I expand to include the modes of thinking used by scientists to discern the truths of the natural world. Mastery of the known facts of any problem must be carefully coupled with a degree of skepticism about previous explanations. Rigorous intellectual integrity demands that one recognizes how self-interest can strap blinders on us to prematurely terminate reasoned inquiry. Chapter 2 will describe how scientists evaluate, test, reject, and modify hypotheses. Furthermore, it will list warning signs that allow the detection of sterile hypotheses that must be molded into useful ones.

But this is not primarily a book on the theory of the scientific method; I have read several of these and find them dry as dust and less gripping than a telephone directory. By contrast, the actual process of scientific discovery is a roller coaster

How the Great Scientists Reasoned. DOI: http://dx.doi.org/10.1016/B978-0-12-398498-2.00001-1

of exhilarating highs and deflating valleys. It is as delightfully absorbing, as complex, and as profoundly human as falling in love. To properly focus attention on these creative aspects, the majority of the book is comprised of biographical sketches that illustrate important aspects of the methods of scientific discovery. They are not complete biographies and are not intended to illuminate the full character of each scientist because they are focused on the thought processes that led each in his intellectual journey. Sketches of eight scientists intended to illuminate different and important facets of the process of scientific discovery are included.

As I began to compile these stories I found the keenest pleasure in reliving each moment of discovery, each "eureka" flash of enlightenment. Many scientists have enjoyed such experiences—times when all the conflicting pieces of the puzzle neatly slide together to make a beautifully coherent whole. A mixture of emotions can overpower the mind at such times: reverence for the beauty and harmony of nature, pride of achievement for unraveling such a knot, and gratitude for the privilege of being the first to appreciate and enjoy such a mystery. Just for a moment, the researcher has heard the strains of the music of the spheres, as chords hidden since the formation of the universe resonate just for him or her.

Just as the spy novelist revels in creating a fantasy world of dapper but deadly secret agents or the author of bodice-ripping romances savors a heroine's moments of passion, I have delighted in re-creating these moments of discovery. So in the end I have written this book to relive and to share these all too rare instances so precious to the scientist and so useful to the rest of humanity. I hope you will enjoy them too.

My choice of subjects is not as capricious as it might appear. I apologize that the selection does not meet the demands of political correctness, as all eight subjects are white European men. Two of them are chemists, one an explorer, and the remaining five physicists. These eight men, however, were chosen because they illustrate well the different modes and problems of scientific thought.

Chapter 3 describes the scientific aspects of the career of Christopher Columbus, who most people would classify as an explorer rather than a scientist. However, chronicling the struggles of this giant of discovery in a century groping its way out of the medieval and into the Renaissance can convey useful lessons about scientific thinking. Because Columbus could not amend his original hypothesis that he would reach Japan by sailing 2500 miles to the west, despite many contrary facts he uncovered during his four voyages, his valid claim to the discovery of two continents new to European civilization was weakened and his life ended in frustration.

Chapters 4 to 6 describe the discoveries of the experimental scientists Priestly, Lavoisier, Faraday, and Röntgen. These giants beautifully exemplify the productive modes of scientific thought described in Chapter 2. The demise of phlogiston theory described in Chapter 4 is another cautionary tale for all scientists illustrating the lengths to which the human mind can go to retain very flawed conventional wisdom. The researches of Faraday described in Chapter 5 are glimpses into the mind and procedures of a peerless laboratory scientist. Faraday's discoveries were spectacular, but this genius at synthesizing information did commit a misstep when he tried to merge the forces of electromagnetism and gravitation. Röntgen's discovery of X-rays described in Chapter 6 is a tale not told often enough, illustrating how a careful and thoughtful observer can discover what Mother Nature had been shouting for years to a host of scientists who did not listen carefully enough.

Each of these four experimentalists showed a fertile skepticism that enabled them to push beyond the boundaries set by the past. Each was capable of communing with Nature in their laboratories: listening, testing, discarding, imagining, and formulating more accurate hypotheses.

Nevertheless, each of them might be criticized, for they all made errors of fact and judgment. The demands of truth are very high, the laws of nature can be intimidatingly complex, and human ego makes our minds frail. However, even though none achieved perfection, these giants made monumental contributions to human welfare.

In Chapters 7 to 9, we vault into the twentieth century. In the first two decades of this new age, the increasing scope and specialization of the scientific community created three outstanding theorists: Planck, Einstein, and Bohr. They fed off the growing vitality of the scientific enterprise, surveying the most exciting experiments, picking out the results that could not fit the classical understanding of physics, and assembling the complex framework of quantum mechanics to rationalize the hitherto inexplicable. These giants knew each other, and both Bohr and Einstein expanded Planck's original quantum hypothesis in ways that he initially found troubling. The foundations these three men built for the new quantum theory led to the most bizarre, sophisticated, and successful theory conceived by the mind of man.

I originally hoped that this book could be written with only a bare minimum of equations. However, I found that adding some mathematics to Chapters 7 to 9 enriched their story considerably. Readers not having the background to understand these equations are encouraged not to despair, because I have tried very hard to incorporate the meaning underlying the mathematics within the text.

In the short conclusions chapter, I tie together some of the disparate threads of scientific thinking and apply them to our current world.

After you have read this book, you would have traced the reasoning of some of the most independent, careful, and concise thinkers who have ever lived as they worked their way through very knotty conundrums. Each had to carefully winnow valid observations and data from a cloud of confusion and blunders. If you can assimilate, even to a small extent, the habits of thought they cultivated, then you will find that solving the problems of your own personal life—who to vote for, what investment decisions to make, which medical procedures to trust, what career to prepare for, and so on—will be founded on a more rational basis.

Good luck!

2 Elements of Scientific Thinking: Skepticism, Careful Reasoning, and Exhaustive Evaluation Are All Vital

2.1 Science Is Universal

Suppose you travel to an exotic location like Indonesia, Israel, or Japan. Compared to the United States, the languages are completely different. The predominant religions have little in common. The governments are structured differently, and the political beliefs of the average citizen are widely divergent from those of your homeland. Family life will have a different flavor. Yet if you enter a local university in each of these countries and examine the content taught in a department of chemistry or biology or physics, you will find little variation in point of view or subject matter.

This is not just a question of a university classroom being a sophisticated locale. Even within the United States, there are broad differences in the content and teaching in the fields of religion, psychology, English literature, and political science, but very little divergence of subject matter in the teaching of chemistry or geology or physics.

In the pages to follow, I will argue that this broad agreement on what is accepted scientific theory is an affirmation that scientists have devised effective means for winnowing out errors and retaining theories that contain a valid description of Nature. The key element that has allowed this progress is that the hard sciences must agree with experimental and observational reality.

To capsulize and oversimplify the extreme case of the difference between scientists and politicians: the scientist is accustomed to seeing her ideas proved wrong and subsequently modifying them, whereas a politician never admits that her policies are wrong and carefully minimizes and disguises any necessary alterations of an original position.

2.2 Maintaining a Critical Attitude

We all have learned from painful experience how often people prevaricate and dissemble. Blind reliance that others will tell us the truth is a demonstrably risky policy.

How the Great Scientists Reasoned. DOI: http://dx.doi.org/10.1016/B978-0-12-398498-2.00002-3

More frustrating yet, people who really love us may not always tell us the truth. Even those who sincerely have our best interest at heart may distort the simple truth, deny events that actually happened, or fabricate untruths that they feel are important for us to believe. Highly respected world leaders have been known to deliberately lie to their countrymen.

Sometimes these distortions are forgivable. You may scan all of Abraham Lincoln's writings about the American Civil War and you will never read that the war was about slavery. It was about preserving the Union. Perhaps in Lincoln's mind the two were inextricably intertwined, but a clear, simple description of this war would have to include the observation that in the Northern states the idea of slavery had become so onerous that allowing its practice in the United States of America had become unacceptable. We can understand that during the entire war, Lincoln cherished the hope that the hostilities could be ended on terms acceptable to the South and that soft-pedaling the slavery issue would give him vital political flexibility. But he was not relating the clear simple truth that the American Civil War was about slavery. The fact is that Lincoln was more committed to something he felt was more important than truth: preserving the Union and conserving the lives of American soldiers.

A more sinister compromise with truth is visible in the responses of the leaders of the Catholic Church to the problem of pedophile priests. Although I am not a Catholic, I am convinced that the Church is staffed and led by well-meaning and generally trustworthy men who believe that their religious efforts are vital to a suffering and sinful world. But the problem is that their primary commitment is to the institution of the Church itself. And for many who climb in the Church hierarchy, commitment to the institution of the Church can become more important than commitment to the truth. After many years of cover-ups, we have learned that bishops, cardinals, and even popes were far less committed to preserving the sanctity of our children than protecting the reputation of their church.

Despite the prevalence of mendacity in our human relationships, science is itself a search for truth and can thrive only in an atmosphere of forthrightness and candor.

In most scientific areas, at least, truth can be suppressed only so long. There is a natural vitality to truth; it is like a mighty Mississippi whose rolling current daily scours and cleanses its bed. In times of flood, its unstoppable torrents wash away deep-rooted misconceptions, superstitions, and distortions. Truth is nurtured by unbiased and judicious thought, and clarity and straightforwardness make truths powerful and easy to test.

However, the self-interested prevarications of others are not nearly as dangerous to clear thinking as the distortions we fabricate for ourselves. Ambition, pride, or stubbornness can make us blind to our own follies and the most obvious of truths. Chapter 3 on Christopher Columbus relates how an intelligent and perceptive man could ignore obvious truths long after they were accepted by lesser intellects. Clear thinking definitely has an ethical component that ambition tries to strangle.

2.2.1 Reasonable Skepticism

Skepticism has an even more important role to play than helping us unravel deliberate lies; it arms us to question the honest errors that are so lamentably prevalent in the books, newspapers, electronic media, and the well-meaning words we hear every day. A careful researcher quickly learns that even the scientific literature, despite heroic efforts of editors, reviewers, and authors, is filled with oversimplifications, overgeneralizations, and mistakes.

A healthy skepticism is the necessary forerunner of truth. The fields cannot be sown with productive crops unless the worthless weeds are first uprooted.

2.2.2 Respect for the Truth

A deep and committed respect for the truth—whatever it may be and whomever it may advantage—is the most productive frame of mind for evaluating evidence.

Very frequently, however, knowledgeable researchers slip into the mode of thought that they are trying to find evidence to prove X. If you are in the laboratory doing experiments attempting to prove X and Nature keeps responding that Y is correct, then you are all too likely to ignore or underweigh serious clues. Retaining an unbiased and calm mind will more efficiently allow you to accept valid new evidence as you uncover it.

But here's the catch: It is virtually impossible to remain unbiased in a field of research where you have been active for several years. You have doubtless published papers where you suggested that Z is true. You have addressed your colleagues at scientific conferences and seminars and have marshaled strong evidence for Z. So when a novel series of experiments yield evidence that Z is a rather simplistic hypothesis and should be abandoned, you might feel some embarrassment. Acclimating yourself to the new facts therefore takes extra time and energy, and the original commitment to Z, once only a working hypothesis, has become a stumbling block.

It is best to remember that the researcher is just asking questions; it is Nature's job to provide the answers. The researcher's toughest job is to listen. Chapter 6 on Röntgen demonstrates just how effective a single researcher, who carefully absorbs unexpected evidence, can actually be.

2.3 Reasoning

There are two modes of reasoning, deduction and induction, that allow us to extract general truths from known principles, observations, or experimental data. Even though one might hope such reasoning might be sufficient to convince others, gaining converts to new scientific principles is frequently more complicated than simple persuasion. A twentieth-century reinterpretation has introduced the notion of *paradigm shifts* to describe the sociology of the acceptance of scientific revolutions.

2.3.1 Deduction

Deductive reasoning begins with accepted truths and draws logical consequences from them [1]. It requires that you accumulate relevant facts about a problem, carefully weigh and compare them, and deduce a balanced conclusion that will fit all the facts into a consistent framework. This is the breathtakingly beautiful thinking we learn from Euclid's geometry, where an elaborate and useful edifice can be constructed from just a few axioms. This mode of thinking is also familiar to all of us from the detective novel. The hypothesis that Colonel Mustard, acting out of jealous rage, stabbed Miss Scarlet in the garden at 10 p.m. cannot be true if Colonel Mustard was playing bridge in the library with Professor Plum and Mrs. Peacock for the entire evening. Reasoning deductively, you should not have accused Colonel Mustard until you knew his whereabouts at the moment of the murder.

Deduction works most efficiently when the logical framework of the problem is understood. If not, it puts the brakes on the solution of a problem until such facts come in. In many cases, however, a natural phenomenon is so complex that a definitive set of facts is not readily available.

2.3.2 Induction

Inductive reasoning tries to infer general laws from specific observations. For many complex problems, the only guide to gathering the relevant facts is to begin some creative musing about the cause of the phenomenon; this leads to the inductive method. In scientific practice, fact accumulation usually proceeds apace as different researchers bang their heads on a new problem until someone manages to assemble the observations into a partially coherent picture. To the extent that the picture is incomplete or ambiguous, it becomes the hypothesis that future experiments will test and improve.

Science thus generally proceeds by the inductive method of constructing informed surmises about how things happen. These hypotheses may start as little more than guesses or observations with minor intellectual content. For example, "Rain causes rainbows." To make these hypotheses amenable to scientific inquiry, they must be *falsifiable*, that is, one must be able to test them and prove them false if they do not correspond to observation or experiment. For instance, one could disprove the proposition that rain causes rainbows by observing a rainbow on a day when there is no moisture in the vicinity. Falsified hypotheses become the humus from which more accurate hypotheses sprout.

Mythology or religious revelation, as in the Old Testament story of the origin of the rainbow, is not usually falsifiable. According to the book of Genesis, after God destroyed the world by flood, saving a remnant of humanity and animal life through Noah and his ark, he promised that he would not repeat this catastrophe. As Genesis 9:11 tells us, "And I will establish my covenant with you.... I do set my bow in the cloud, and it shall be a token of a covenant between me and the earth... when I bring a cloud over the earth, that the bow shall be seen in the cloud... and the waters shall no more become a flood to destroy all flesh."

It is hard to see how to disprove this assertion. We were not there at the time to observe either God's intervention or communication, so we can neither verify nor refute them. But the text might encourage the skeptical to ask further questions. For instance, are there rainbows in pagan countries where God's promise might be invalid? Are there times of peril when the rainbow of God's encouragement might be more or less intense? The fact that neither of these details has been observed nor does not really refute the hypothesis, or rather "revelation."

Contrast the sterility of trying to evaluate this type of "one-time" myth with the productivity of a hypothesis that does not involve supernatural intervention. For instance, we may refine our first hypothesis that rain causes rainbows to become more specific: Now it might read "Rainbows are created by the refraction of sunlight by airborne water droplets." Even a cursory nontechnical examination of this hypothesis encourages us that we are on the right path; for instance, rainbows are not observed when it is totally overcast and are almost always observed at the end of a rainstorm when there are many water droplets in the air. Moreover, if the air temperature is far too cold to sustain liquid water, *sun dogs*, which usually have a columnar rather than a bow shape, are observed instead.

It is easy to poke fun at the fable of the rainbow, but Chapter 4 on Lavoisier and Priestly will show brilliant scientists struggling with the eighteenth-century dogma of phlogiston, whose adherents stubbornly added layers of complexity each time falsifying data appeared. Even so, the century was known as the "Age of Reason."

Falsification allows one to sort through alternate hypotheses by observation and experiment. Hypotheses that are shown to be false must be altered or discarded. The *doing* of science is ideally the successive refinement of hypotheses; in many cases they converge closer and closer to some physical measurement. Ideally, this convergence persuades all interested parties to adopt the refined hypotheses. This is the goal to which scientists aspire; in our hearts, we see this as the normal state of scientific progress.

A clear example of the convincingly successful application of such thought processes is the discovery of X-rays by Wilhelm Röntgen described in Chapter 6. Röntgen carefully sorted through the possible explanations for his new discovery and convinced the world in one closely reasoned paper that X-rays were electromagnetic waves.

Should we be sobered by the fact that the most profound tool of the scientific method is the falsification of hypotheses, a tool more akin to the wrecking bar than the mason's trowel? What happens when researchers dispose of a naïve hypothesis? No formula or technique is general enough to guide a researcher at this stage. Whether he patches up the old hypothesis, reassembles its components in a more productive way, incorporates new concepts from his own imagination or the experiments of others depends on the exact case. The biographies that follow will illustrate a diversity of approaches.

The development of the scientific method is frequently traced back to Francis Bacon's *Novum Organum* (1620) and René Descartes' *Discourse on Method* (1637) [2]. But Chapter 4 on Lavoisier, Priestly, and phlogiston will make clear how nebulous the understanding of the scientific enterprise was even near the end of the eighteenth century.

If you consider carefully the chain of logic we have just outlined—hypotheses formed, tested, and refuted—it will become clear that such a chain of reasoning can *never* prove a scientific theory completely correct. The last hypothesis standing is just the best approximation that has not been falsified.

In actual scientific practice, some hypotheses seem so complete and unassailable that we dignify them with the term *theory*, as in "general theory of relativity" or "electromagnetic field theory." This honorific gives you a license to use this information without apology, but puts you on notice that if you wish to assail part of the theory you will need extremely compelling justification.

However, a close examination reveals that even the mighty electromagnetic field theory assembled by James Clerk Maxwell, which most physicists consider to be the archetype of beautiful and clear theory, was constructed out of airy and insubstantial metaphors. In a series of papers starting in 1855, Maxwell welded the observations of Coulomb on electrostatics and those of Faraday on electromagnetic induction into a mathematical theory [3]. For physicists since the time of Newton that means presenting a set of equations expressed in terms of differential calculus. Justifying these complex equations was difficult, and Maxwell resorted to analogies such as fluid-filled tubes and rollers and cogs to represent the electromagnetic field. Nevertheless, revised and consolidated versions of Maxwell's equations have survived innumerable tests of their accuracy and have even proved to coexist quite compatibly with Einstein's special theory of relativity.

In writing about the beleaguered financial industry, Nassim Taleb [4] developed a useful metaphor for the difficulties of proving any hypothesis to be correct: the *black swan theory*. Any resident of Eurasia or America would have agreed with the proposition "All swans are white" until the seventeenth century. However, the discovery of Australia and its novel species of black swan refuted this oversimplification. Within a limited region, the hypothesis that all swans are white was useful and consistent. It was just invalid in Australia. All the experience of the most knowledgeable and brilliant biologists was refuted by simple observation on a new continent.

Einstein in speaking about his theories of relativity capsulized it thus: "No amount of experimentation can ever prove me right; a single experiment can prove me wrong."

2.3.3 Paradigm Shifts

Philosopher of science Thomas Kuhn [5] proposed a somewhat different picture of the progression of scientific ideas in 1962. He conceived that the failure of journeyman scientists to solve the day-to-day problems of "normal science" in any area could force a *paradigm shift* to a novel explanation that could supersede the older theories. Kuhn visualized that these paradigm shifts were necessary to create "revolutionary science" as the older theories were supplanted with newer ones, sometimes not even providing more accurate descriptions of Nature. Kuhn's theory opened the floodgates as philosophers began to carefully examine scientific revolutions to discover that there was a lot of art in scientific progress. More extreme philosophers such as Feyerabend noted that the scientific method was really rather a diffuse thing and that scientists sometimes were unfair in their support for their ideas. Might it even be possible that ideas coming from more influential authors or institutions or of more relevance to evolving social or economic power conditions had the inside track on scientific acceptance? It seemed to physicists that our colleagues in the liberal arts, for whom the progression of fashionable, more politically acceptable ideas is as regular as clockwork, showed unseemly haste in amplifying these criticisms.

At least this aspect of the impersonality of science is true: In the current era, when a novel and important theory in the hard sciences is submitted for acceptance, it is quickly jumped on, dissected, reevaluated and extended by other working scientists. In a very few years, it will be sufficiently vetted that the personal aspects of its original discovery become unimportant.

"Who discovered what?" is frequently a much more complex question than the layman might suspect. In writing Chapter 4 on Lavoisier and Priestly, I have tried to convey an appropriate sense of the ambiguity surrounding the "discovery" of oxygen but have simplified the story by trimming out the contributions of Scheele and others. In attempting to unravel what constitutes a paradigm shift, Kuhn [5, p. 55] discusses the philosophical problem surrounding the discovery of oxygen well. The urge, or perhaps the necessity, of simplifying a story tends to aggregate discoveries to the most famous. In Chapter 8 on Einstein, for instance, I found myself praising the subtlety of his analysis of the Brownian motion yet neglecting to mention a nearly identical and simultaneous derivation by Australian W. Sutherland [6]. I likewise could have spent many pages discussing the ideas of Poincaré, who had been developing relativity-like ideas for many years and may have come close to scooping Einstein to the publication of the special theory of relativity [7]. Einstein did not mention Poincaré in his first paper on special relativity and was famously parsimonious in providing references in his papers.

In our modern era, with its rapid electronic communication, abundance of conferences, and large teams of scientists solving major problems together, the ownership of breakthrough ideas is more evanescent than ever before. Scientists love to talk about their work, and the frontiers of research buzz and crackle with newly hatching ideas whose attribution is soon forgotten.

In fact, there are many scientific areas such as particle physics in which the scope and complexity of current research require that large teams of specialists collaborate to solve difficult problems. This approach is carrying some branches of physics and biology far from the methods of Priestley, Faraday, and Röntgen, who toiled self-sufficiently and alone for weeks at a time. In the future, teamwork may be considered just as important a requisite for physics research as it is for soccer. Chapter 9's description of the career of Niels Bohr will celebrate the virtues of teamwork and collaboration.

I myself am convinced that many editors of scientific journals and the reviewers to whom they send submitted papers will give a somewhat more sympathetic reading to a paper submitted from MIT than one from Podunk State University. In fact, there has recently been considerable discussion in the letters column of the physicists' trade journal *Physics Today* as to whether a youthful Einstein submitting revolutionary articles today would be able to get them published. In 1905, the unknown Einstein (Chapter 8) submitted five brilliant papers, among them the special theory of relativity, to the leading physics journal in Germany, *Annalen der Physik.* All were published with little delay, even though Einstein's institution was listed as the Swiss federal patent office, hardly a prestigious address. Perhaps it is relevant that in those days of a small scientific community, the associate editor of *Annalen* was Max Plank, a very gifted scientist, and he was not required to cope with the flood of papers submitted to a leading journal of today.

However, the notion that race, gender, or wealth has a marked effect on the acceptance of the laws of the hard sciences of physics, chemistry, or mathematics is anathema to most working scientists. We shall see in Chapter 4 on phlogiston how Lavoisier's recasting of chemistry in the late eighteenth century met resistance so determined that it was only accepted as the old guard died off, but the physicists I know would sooner confess that their mothers used to turn tricks than assent the validity of this view in modern times. Almost to a one, their view is that science proceeds by successively improving approximations toward some breathtakingly simple, overarching, and shining intrinsic truth. Retrograde steps are not totally unknown but are always short lived.

Our view is that the introduction of quantum mechanics, for example, was dictated by observations and experiments that could not be explained by Newtonian physics. For instance, the radiation emitted by hot bodies (Chapter 7), the photoelectric effect (Chapter 8), and the optical spectra of atoms (Chapter 9) are inexplicable within the framework of Newtonian mechanics and classical electromagnetism. Chapter 9 will describe how thoroughly revolutionary Niels Bohr's description of the quantum mechanical hydrogen atom was and the resistance to its acceptance. Although powerful personalities can and do preserve incorrect ideas in the short term, in the long haul carefully performed measurements can supersede the older ideas. Science refines its ideas in a way that is totally unlike the fashion industry's replacement of lime aqua by honeysuckle orange.

I have had the good fortune to have taken courses taught by Linus Pauling and John Bardeen and to have heard many lectures by Richard Feynman. For these three Nobel Prize winners, there was no immutable scientific method. Courses in scientific method or the philosophy of science were not required or even taught to physicists being educated at Caltech or the University of Illinois. We were taught that what was most economical, tidy, beautiful, and consistent was probably what was true. Results that were clearly in violation of a well-accepted theory were to be warily examined. A very useful theory might be patched up to accommodate a conflicting result, but, just as in your wardrobe, too many patches would require discarding the garment as soon as a better one might be found.

Henry Bauer [8] has pointed out in *Scientific Literacy and the Myth of the Scientific Method* that the working procedures of scientists in different disciplines are highly variable and frequently diverge widely from our model of falsifying hypotheses to create modified hypotheses. In fact, experimental results that falsify revered theories do not lead to immediate rejection of these theories but to more careful experiments, as theory often has more credibility than experiment. A recent example might be the report of faster-than-light neutrino beams from the Large Hadron Collider in Grenoble that stimulated more careful experiments rather than rejection of the special theory of relativity.

The scientific method does not work because some people never make mistakes. It works because truth is more durable than error.

2.4 Evaluating Scientific Hypotheses

Scientists and philosophers have evolved some very useful tools that they return to again and again to help them evaluate, improve, or discard hypotheses.

2.4.1 *Ockham's Razor*

A principle often attributed to a fourteenth-century friar, William of Ockham [9], can be useful in sorting through hypotheses. It is frequently rendered "All things being equal, the simplest solution tends to be the best one." This idea is economical and appealing; it has proved to be a good general rule of thought.

Einstein rendered this idea with crystalline clarity: "Make things as simple as possible, but no simpler." We will tell a cautionary tale in Chapter 5 about Michael Faraday's unsuccessful attempt to consolidate electrodynamics and gravitation. In terms of the Einstein quote above, Faraday was trying to make gravity simpler than it really is.

Our rainbow hypothesis has several elements. Can we eliminate one? We need both water droplets and sunlight (or at least a directed light source), and our preliminary observations show that both are necessary.

The entire theory of evolution is a gigantic Ockham's razor simplification. You may choose to believe that a Creator established thousands of similar species differing only in location and food supply, or you may accept that natural selection is operating continuously to modify and produce new species better adapted to the inevitable changes in habitats.

Ockham's razor also dictates, for instance, the way to deal with many hypotheses involving extraterrestrials. One must simply decide whether the Nazca Indians of Peru developed excellent skills at drawing figures and straight lines hundreds of meters long in the desert, or they were assisted by aliens who otherwise left no trace of their visits. Did the Egyptian pharaohs develop means of siting and constructing the pyramids, or did aliens likewise teach them, leaving no additional clues of their benign help? Although visits by aliens can never be totally ruled out, a sensible weighting of these alternatives forces us to temporarily discard such unlikely events pending the discovery of convincing artifacts.

There is a serious complication that sometimes makes it difficult to apply Ockham's razor, because as we expand our knowledge and the phenomena we wish to explain in any field, the theories that describe its behavior inevitably become more complicated. Heat is conducted across insulating solids, for example, through propagating waves called *phonons* that vibrate their constituent atoms. This happens in metals, too, but a larger portion of thermal conductivity comes from the mobile electrons that all conducting metals but no insulating solids contain. One mode of heat conduction has, unfortunately but legitimately, multiplied into two.

Chapter 7 on Planck will describe the strenuous efforts of a conscientious scientist to avoid adding a new hypothesis to the science of thermodynamics: the idea that an oscillator can only radiate or absorb discrete units of energy. Planck only let himself be convinced by the marked success of his theoretical expression describing the light radiated by hot bodies. Although many other scientists were slow to follow Planck's lead, Einstein was able to broaden the scope of this hypothesis by applying it to light, whereas Bohr applied it to electronic energies within the atom.

2.4.2 Quantitative Evaluation

The true clincher in evaluating a hypothesis is quantitative prediction. It is the supreme test that indicates how accurately the truth is being approached. By studying the diffraction and reflection of light beams passing through water-filled spherical vessels, Descartes [10] was able to show in 1647 that only the droplets aligned at an angle near 42° to a line through the sun and our eyes can scatter rainbow light into our eyes, thus accounting for the observed arc shape and angular disposition of the rainbow (Figure 2.1). A few years later, Isaac Newton showed that sunlight is comprised of many colors and that in the refraction and reflection process red light will be displaced through a smaller angle than blue, accounting for the color dispersion observed. Later workers have filled in the picture more completely, accounting

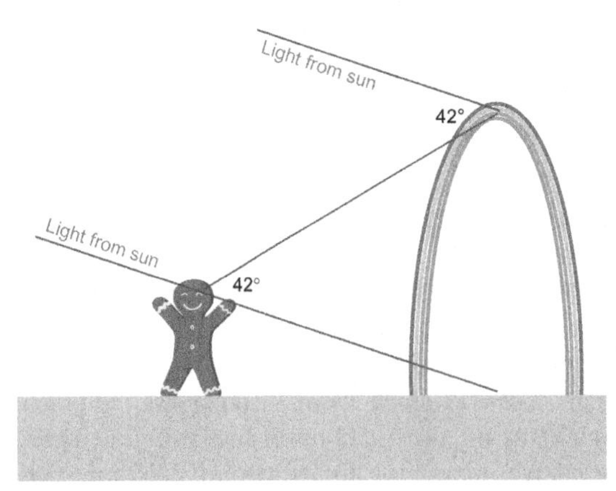

Figure 2.1 Geometry of the rainbow. The rainbow. Sunlight refracts as it enters and leaves each droplet, forming a circular arc oriented about 42° from a line through the sun and the observer. Droplets at all distances from the gingerbread boy contribute to the rainbow; its position in the figure just indicates the angle at which he perceives it.

for many more complex features associated with the scattering of light by water droplets. This quantitative agreement makes it pretty convincing that we understand the natural phenomena of rainbows well. In general, quantitative agreement thus can tell us that one theory works better than another and lets us know just how accurately we can describe the phenomenon.

Once we begin thinking in a quantitative way, estimation becomes an indispensable tool of thought. Quantitative thinking can often be applied in a way that guides logical thinking, even when hard information about a poorly defined problem is not available.

For example, suppose you have a deadline such as a movie that begins in 1 hour. You peek into an unknown but attractive restaurant and wonder if you can order and consume a simple meal in less than 1 h.

Here is a statistical method for making a useful guess as to what the total time you will spend seated at your table might be. Look around the restaurant and determine at what fraction of tables people are eating. Sipping a drink or playing with the cold remains of a lunch does not count. For instance, you might be able to scan six tables without looking embarrassingly inquisitive. Of those six, two are actively eating. On average then, you might expect to be seated for three times the length of time required to consume your meal. If this is a simple one-course lunch, then you might anticipate being seated for less than 1 h (3×15 min = 45 min total). It is a reasonable bet you can make your movie without alienating the restaurant staff.

For a multicourse meal, you would have to increase the actual eating time, perhaps to 35 min if you include a salad and dessert.

This method is especially useful if you want to screen out restaurants with very poor service. Suppose you peek in a restaurant and see no one eating at the eight tables you observe. Generally, that means you are in France. However, if you do happen to be in the United States, it means that service has broken down and you can expect a very long wait.

Much of the bread and butter work in modern science and engineering describes phenomena like restaurant service that cannot be pinned down to good accuracy. One example of critical importance to those of us who ride in airplanes and automobiles would be determining the fatigue life of metallic alloys. It is vital that the microscopic damage accumulating in a Boeing 747 during each flex of its gigantic wing not aggregate into cracks that might lead to catastrophic rupture [11]. Making accurate measurement of an aluminum alloy's fatigue life and ensuring that this is representative of every structural portion of wings manufactured in the future is so difficult that the only responsible approach is to allow a very large margin of safety, meaning *orders of magnitude*, not just a few extra percent.

In a somewhat more accurate venue, studies of the physical properties of solids, scientists proudly publish papers comparing parameters like the heat capacity or light absorption of new materials that agree with established theories to within a few percent. Errors of this size might easily originate from experimental problems such as impurities, imperfections in crystal structure, or surface roughness.

But physicists and chemists have most generously lavished their attention on precisely determining the fundamental constants of Nature that recur frequently in scientific calculations. Chapter 8 on Einstein's work that follows will describe his efforts to accurately evaluate Avogadro's number, which is proportional to the number of molecules in a gram of a pure substance; it is related to the reciprocal of the weight of a molecule. Einstein's efforts have been substantially improved by the discovery of X-ray diffraction, a process that determines the spacing between atoms in a crystal very precisely, so that Avogadro's number is now believed to be

$$N = 6.02214129 \times 10^{23} \text{ molecules per gram molecule,}$$

to a relative accuracy of 44 parts per billion (10^{12}).[1]

The elementary charge on an electron or proton is determined by weighing the material deposited in an electrochemical cell. This determination is reliant on Avogadro's number, so it is just marginally more accurate than Avogadro's number itself:

$$e = 1.602176565 \times 10^{-19} \text{ Coulombs,}$$

to a relative accuracy of 22 parts per billion.

A vast improvement in accuracy can be obtained with a purely spectroscopic measurement such as the minimum energy required to eject an electron from the most tightly bound orbit in a hydrogen atom, the Rydberg. This depends only on an accurate measurement of the frequency of ultraviolet light, something modern electronics can do with superb accuracy. So current measurements show that the wave number (the reciprocal of the wavelength) of the Rydberg is

$$R = 10973731.568539 \text{ per meter,}$$

to a relative accuracy of five parts per trillion (10^{12}).

It is important to understand that these numbers are not totally independent, so that if there were sizable errors in the claimed values or accuracies in any of them, alarm bells would be going off in the lengthy list of physical constants that the National Institute of Standards and Technology in the United States and other institutions abroad maintain. These numbers are not the last word but will be improved year after year. Efforts to improve them will only cease with the end of our scientific civilization.

If you are seeking perfect, changeless accuracy or explanations that are never refined or updated, then you must enter the realm of faith and religion. Your major problem then will be to decide *which* religion to entrust with the awesome responsibility of being eternally correct.

[1] 2011, NIST.gov.

Chapter 9 on Bohr will describe how he was able to convince the world of the correctness of his unprecedented quantum mechanical model of the hydrogen atom and helium ion by calculating the hitherto inexplicable frequencies of light emitted by these excited atoms to an accuracy of about two parts in a million.

Ideally, the gold standard of quantitative evaluation would be a true *prediction*: that is, the calculation of a physical parameter *before* it has been measured. This is a much more convincing verification of a theory than a *postdiction*, or agreement between a new theory and a property that has already been measured. Because new theories arise from disagreement between existing theories and measurement, predictions are much rarer than postdictions. We shall see that Einstein's general theory of relativity actually predicted the bending of starlight by the sun, a previously unknown effect, although the experimental verification may have been premature in that it was unduly influenced by the theoretical prediction.

2.4.3 Verification by Others

In principle, the scientific method protects against error or fraud by confirmation of any significant discovery by other outside groups. An exciting new claim will motivate the ambition of others to confirm or disprove the results; either way it stimulates excitement at the frontiers of science.

This system worked perfectly after a report describing a new class of high temperature superconducting ceramics in 1986. Two scientists from Switzerland's IBM laboratories, Karl Müller and Johannes Bednorz [12], observed that a complex material containing copper crystallized in the *perovskite* structure could remain superconducting at a record high of 30 K, or 30 degrees Celsius, above absolute zero. The technological implications of this discovery were exciting, nurturing the hope that superconductors might be developed that could transmit electricity without loss while only requiring relatively inexpensive cooling by liquid nitrogen, which boils at 77 K. Measuring the temperature at which a solid's resistivity drops below detectable limits is a fairly standard experiment, and the ability to make this measurement is widespread. Within weeks, the initial claim was duplicated at laboratories around the world, and many new and even superior materials were quickly developed. Incidentally, this class of superconductors cannot be tidily wedged into the previously existing Bardeen–Cooper–Schrieffer theory of superconductivity, so what was once a neatly explained field still lies in disarray.

Chapter 8 will describe how the young Einstein kicked off his magnificent career by devising several independent methods of calculating the size of the atom. Einstein thus verified his own work and added overwhelming credibility to the atomic theory.

The system of verification by others did not quickly resolve the Stanley Pons and Martin Fleischmann *cold fusion* controversy in 1989 [13]. These researchers claimed that a palladium cathode in an electrochemical cell could be saturated with such a high pressure of the hydrogen isotope deuterium that the barrier to nuclear fusion could be surmounted and helium formed. This event would be expected to release large quantities of heat, abundant free neutrons, and the radioactive isotope tritium. Their original report was that so much extra heat occurred in their electrochemical cell that it could only be attributed to nuclear fusion. In this case, verification of the excess energy depended on a tricky measurement: carefully determining the energy balance of an electrochemical cell as it dissociates water and evolves hydrogen and oxygen gas. The electrochemists who could perform these experiments accurately were generally not well equipped to carefully measure the flux of neutrons that would have been emitted if nuclear fusion were taking place. Furthermore, passions were inflamed when the administration of the University of Utah became convinced that a legitimate discovery was being suppressed because of lack of respect for its bucolic origin. These problems added enough fuel for the controversy to rage for years. Diehards still meet and nurture hope that cold fusion may power our world in the future, but their work carries more caveats than before. As cold fusion languished, its defenders, trying desperately to retain credibility, recycled some familiar themes common to the death throes of other hypotheses: that the reaction works only "sometimes" or depends on "proper conditioning" of the electrochemical cell.

For research that demands large-scale statistical studies, verification by other laboratories will probably be slow. Checking the potency or safety of a new drug might demand a clinical trial lasting many years. And in a much more political and even softer science, economics, the effectiveness of a Keynesian stimulus is still under vigorous debate almost a century after its original formulation!

2.4.4 Statistics: Correlation and Causation

Imagine that you are an advocate for the benefits of marriage. Among your acquaintance, you observe that those who are married tend to live longer. You wish to test this hypothesis, so you acquire a large database and sort through it. A quick tabulation shows you that people who are married live longer and even report happier lives.

At this stage, you must take a very sophisticated look at the data because you may be a victim of *confounding*. What you wanted to show was that a couple who marries will live longer on average than one that forgoes marriage, i.e., that the act of marriage in itself extends life. But what about those who are incarcerated, seriously ill, abuse drugs, or are destitute? These groups are all less likely to marry and less long lived than the general population. Your unmarried group is more densely populated with such individuals and its members will consequently live shorter lives than the members of the married group. If you wish to statistically buttress your hypothesis, you must carefully control for these and other *confounding* effects.

Drug manufacturers solve this problem by randomly selecting members of their drug study into a control group that will receive a pill containing only sugar. Until we can convince people to marry or forgo marriage based on their membership in a randomly chosen control group, the exact benefits of marriage will remain subject to debate.

Because they are also subject to abundant confounding factors, it is hardly surprising that those who intend to profit by magnifying or minimizing statistical differences between African Americans, Hispanics, and whites; rich and poor; and men and women and can arrive at such conflicting conclusions.

Cavalier abuse of statistics is the staple of the type of soft science we read in newspaper feature pages: Will taking zinc pills help you avoid bowel cancer? Does the consumption of avocados shield you from Alzheimer's? Perhaps or perhaps not, but it would only be prudent to avoid potentially dangerous changes in your personal life based on superficial statistical studies.

2.4.5 Statistics: The Indeterminacy of the Small

Do nearby power lines cause cancer? Do cell phones cause lymphoma?

Well, we know for sure that very-high-frequency radiation such as X-rays and energetic atomic particles can cause cancer. Particles and rays having energies above hundreds of electron volts can crash into bodily organs and cause damage to functioning parts of our cells, because the energies that hold molecules together are in the range of just a few eV. But the energy of a quantum of 60-cycle power line radiation is less than 10^{-16} eV. Cell phone frequencies are much higher, so a quantum of cell phone radiation could have an energy as large as 10^{-5} eV. The latter is in the range of quanta that carry heat, so photons of this energy are constantly bombarding us. Excessive heat, of course, can be dangerous, but we are very familiar with its symptoms. There is no mechanism currently known that suggests that quanta of this small energy can damage tissues or cells in the body except by heating them.

Nevertheless, because this could conceivably be an appreciable public health hazard, many scientists have decided to be scrupulously careful and gather data comparing the health of those who live close to power lines with those who live far away. Many such studies have been performed, with the results following a familiar pattern [13]. Overall, the health of those who live near power lines proves not to be inferior to those who live far away. However, the researchers who have invested their energy in such a study naturally comb through the data to see if there are any specific cancers or health risks that are associated with exposure to power lines. Among the many types of cancer, some of them extremely rare, and the laws of statistics alone will dictate that, even if no real danger exists, a few will show incidence above the statistically expected amount.

"Voila!" says Professor X, "We suspect that power lines cause a relatively rare type of brain tumor, but since the statistics are too scanty, further research must be done. By the way, I'm available to do it!"

2.4.6 Careful Definition

Several of the greatest successes in twentieth-century physics occurred because perceptive physicists carefully defined and sedulously evaluated concepts that previously seemed obvious but actually required rather deep thought.

The special theory of relativity is built on careful definition of what *simultaneous* means. Within the assumptions of special relativity, events that are simultaneous to a viewer at rest will generally not be simultaneous for an observer whizzing by at high velocity. As we will see in Chapter 8 on Einstein, this and other bizarre effects stem from the fact that the velocity of light is unaffected by the motion of its source or any observer, no matter how fast either is moving.

The development of quantum mechanics required scientists to spend decades carefully sorting through ideas such as how accurately electrons may be located. Large masses such as billiard balls and marbles never confuse their identities or locations; one can be certain that their positions and velocities are clearly definable and never confused with nearby objects. Quantum mechanics, however, represents electrons by waves. When two electrons approach each other, the wave function representing both will overlap, and one can no longer determine with certainty which electron is which. This problem was a key stumbling block for acceptance of quantum theory, and vestiges of it are still being clarified and debated today. Most physicists accept the resolution that only by measurement can an electron's properties such as location, spin, or momentum be completely determined. So carefully thinking about what "measurement" is has developed as a key feature of quantum mechanics. Furthermore, the very mathematics of representing a particle by a superposition of waves leads to the Heisenberg uncertainty principle that the momentum and position of a subatomic particle cannot in principle be precisely and simultaneously determined.

2.5 Science at the Frontier

The thoughtful reader will by now be wondering about the tension between Ockham's razor simplification and the manifold phenomena that Nature has surrounded us with. How will we know when a theory simply becomes so overloaded with assumptions that it is useless and should be jettisoned? Chapter 4 on Lavoisier and phlogiston will describe the death throes of a clunky century-old theory of combustion and its replacement by the economical and tidy identification of oxygen that ultimately became the bedrock of modern chemistry. After brilliant scientists have completed the painstaking work of clarifying such complex phenomena, the casual reader will wonder why it took so long. But in the thicket of conflicting and sometimes incorrect observations, personal agendas and confusing complexity, groping toward a consistent solution can be surpassingly difficult.

2.5.1 When Good Theories Become Ugly

Of course, we have our dirty linen even now. And modern cosmology is a lesson in just how ugly a promising theory can become when hypotheses must be modified many times to reflect accumulating observational data. In the 1930s, the fabrication of the 100-inch Hooker reflecting telescope at Mt. Wilson Observatory allowed Edwin Hubble to carefully measure the displacement or redshift of key spectral lines in light coming from distant galaxies. Theory clearly shows, and Hubble independently verified, that the larger this redshift is, the greater is the velocity at which this source of this light is moving away from us. Over a period of 20 years or so, Hubble [14] carefully accumulated measurements showing that the universe was very much bigger than previously expected, and almost all of the galaxies in the universe were moving away from us. The data seamlessly led to the appealing picture of an initial "big bang" occurring about 14 billion years ago from which all the matter in the universe (except in some nearby galaxies) is moving on trajectories away from us. That is not to say that we are at the point of the original bang but that space unfolded from the bang in a way that makes every point in our universe move away from every other point. (Think of points on the surface of an inflating balloon.)

In 1964, the discovery that the universe is bathed in a sea of microwave energy gave startling confirmation to the theory because the energy of these microwaves is consistent with light that should have been emitted when free electrons and nuclei cooled sufficiently to combine into atoms about a third of a million years after the Big Bang. Astronomers continued measurements to refine this successful explanation of the birth of the universe with renewed dedication.

However, starting in the 1930s, but with more convincing documentation in the early 1970s, weird data started to accumulate. Measurements of the velocities of outlying stars orbiting nearby galaxies yielded rotational velocities much larger than could be accounted for by the masses of the visible stars in the galaxies. Somehow, space around these galaxies must be filled with massive amounts of an unknown material that did not shine like stars or absorb light like dust. The stronger gravitational pull from this extra mass increased the velocities of stars orbiting the galaxies in the same way that swinging a bucket about your body faster makes it pull harder on your hand. Painstaking observations over many galaxies showed that this *dark matter* must comprise more than 80% of the gravitational matter in the universe, but its true nature currently remains in the realm of speculation.

In the 1980s, cosmologists began developing a theory called *cosmic inflation* that hypothesized that immediately after the Big Bang the universe expanded with unexpected rapidity, an idea thought to be necessary for the development of the large-scale structures such as stars and galaxies that are important features of the universe we inhabit. The complexities of this idea, in particular what started inflation and what caused it to cease, are still being debated today.

It got even worse. Improved redshift measurements on galaxies very far away convincingly showed that the universe is now expanding more rapidly than ever—and still accelerating. This strikes at the heart of the original Big Bang model that all matter received a tremendous impulse at the instant of the Big Bang and has been continuously slowing down due to the gravitational attraction of all the matter created. The dark matter we just hypothesized should be further slowing down the speeding galaxies, not speeding them up. The tentative fix called *dark energy* is hypothesized to accelerate matter more and more rapidly as the expansion of our universe continues. The nature of dark energy is even more mysterious than dark

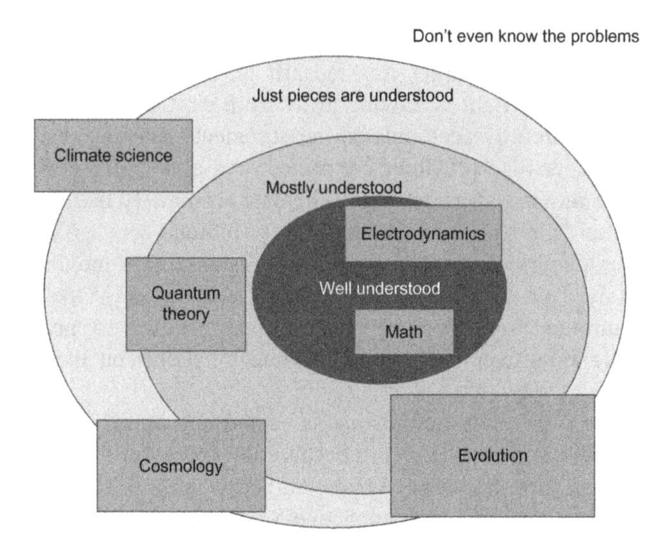

Figure 2.2 Representation of the degree to which some current theories are understood.

matter: It may be built into the very nature of space itself, mandating that as space expands out of the Big Bang it accelerates the matter within it.

In the end, a compelling picture that began with matter speeding away from a primal explosion has had to be cobbled up with three new and startling hypotheses to explain the more detailed data accumulated in the last 40 years: dark matter, inflation, and dark energy. For most theories, these elaborations would definitely be deal breakers and the search would be on for a whole new theoretical framework. But perhaps we have to show a little patience here. After all, this is the theory of the universe.

2.5.2 Stuff That Just Does Not Fit

There are many other areas having observations that do not quite fit the otherwise trusted theories, even after exhaustive investigation by many researchers. Some problems may just be flawed measurements, but others may be indications that current theories are oversimplified and thus act as catalysts for new theories. Einstein's general theory of relativity (Chapter 8) was given a boost in credibility in 1916 by its ability to calculate (postdict) the anomalously large advance of the perihelion of the planet Mercury, which was inconsistent with Newton's theory of gravitation.

Figure 2.2 shows a way of visualizing the different types of scientific knowledge. A solid core represents the things we really are confident about—things like Euclidian plane geometry, which is unlikely to ever prove false. As we move out from this solid core, we pass through the trusted theories of physics such as Maxwell's electrodynamics and special relativity's kinematics.

On the outside of this solid core are theories such as Darwin's evolution whose broad outlines now seem convincingly correct but whose scope is so inclusive that new observations continually test it. A flood of new knowledge in biology will refine this complex and ambitious theory and flesh out the molecular details about how mutation and natural selection really operate, or else they will alter or even supplant it.

By the way, those for whom the theory of evolution is religiously unpalatable can take heart that they are not really opposing a unified phalanx of defenders. Science does not work that way. Any brilliant, eager beaver young biologist would be very happy to drive a stake into the heart of the theory of evolution if she could because it would enhance her reputation and professional standing. She would have to meet a demanding standard of proof to convince colleagues, though. Note that this idea is diametrically opposed to the mindset of organized religion where custom exalts faith and actively squelches questioning.

Even further out on the margins of science are areas with big political ramifications such as climate science's theory that human-produced CO_2 is now significantly warming earth. Although it seems unambiguous that the human combustion of fossil fuels has increased atmospheric CO_2 from near 270 ppm (parts of CO_2/million, by volume) to more than 380 ppm over the last 250 years, calculating by how much earth's temperature will rise because of this is extremely complex and depends on many assumptions, for instance, about how cloud cover will be affected. The data implying a recent temperature increase are noisy and the period is short; moreover, it does not convincingly track the measured CO_2 increase. Politics is now deeply enmeshed in this theory; if you seek to fund research questioning the extent of human-caused global warming, you are extremely unlikely to receive aid from government agencies in the United States or Europe. These are not good conditions for an unbiased search for truth.

On the outskirts of these and many other theories, working scientists camp and earn their daily bread.

References

[1] I. Asimov, Asimov's New Guide to Science, Basic Books, New York, NY, 1984, pp. 9–15.

[2] F. Bacon, The new organon, in: L. Jardin (Ed.), Cambridge Texts in the History of Philosophy, Cambridge University Press, Cambridge, 2000.R. Descartes, *Discourse on method* (1627), in: J. Cottingham, et al. (Eds.), The Philosophical Writings of Descartes, Cambridge University Press, Cambridge, 1988.

[3] W.D. Niven, The Scientific Papers of James Clerk Maxwell, Dover, New York, NY, 1952.

[4] N. Taleb, The Black Swan: The Impact of the Highly Improbable, Dover, New York, NY, 2010.

[5] T.S. Kuhn, The Structure of Scientific Revolutions, third ed., University of Chicago Press, Chicago, IL, 1996.

[6] W. Sutherland, A Dynamical Theory for Non-Electrolytes and the Molecular Mass of Albumin, Phil. Mag. 9 905 781–785.
[7] A. Pais, "Subtle is the Lord...": The Science and the Life of Albert Einstein, Clarendon Press, Oxford, 1982, p. 128.
[8] H.H. Bauer, Scientific Literacy and the Myth of the Scientific Method, University of Illinois Press, Urbana, IL, 1992.
[9] Ockham's Razor, Encyclopaedia Britannica Online, 2010.
[10] R. Descartes, Discourse on method (1627), in: J. Cottingham (Ed.), The Philosophical Writings of Descartes, Cambridge University Press, Cambridge, 1988.
[11] W.D. Callister, Materials Science and Engineering, An Introduction, third ed., John Wiley & Sons, New York, NY, 1994, p. 89.
[12] G.J. Bednorz and K.A. Mueller, Possible high Tc superconductivity in the Ba-La-Cu-O system, *Zeitschrift für Physik* B 64 (1986) 189–193.
[13] R.L. Park, Voodoo Science: The Road from Foolishness to Fraud, Oxford, New York, NY, 2000, pp. 11–27, 140–161.
[14] This story is told well by Hawking SW, The Universe in a Nutshell, Bantam, New York, 2001.

Bibliography

T.S. Kuhn, The Structure of Scientific Revolutions, Third Edition, University of Chicago Press, Chicago, 1996.
This work frames the modern debate on the scientific method.
N. Taleb, The Black Swan: The Impact of the Highly Improbable, Dover, New York, NY, 2010.
More finance than science, but valuable.
R.L. Park, Voodoo Science: The Road from Foolishness to Fraud, Oxford, New York, NY, 2000.
A great read!

3 Christopher Columbus and the Discovery of the "Indies": It Can Be Disastrous to Stubbornly Refuse to Recognize That You Have Falsified Your Own Hypothesis

A working hypothesis is frequently not much better than a guess, but it is a guess that should be altered and improved as new and germane data are acquired. We will see how Christopher Columbus sought financial support for his voyage to "the Indies" with a deeply flawed hypothesis vastly underestimating the length of such a voyage. There was no room in Columbus's hypothesis for new continents, and he would not settle for such an outcome. Columbus stubbornly went to his grave insisting that the Caribbean islands he discovered on his first voyage were the Indies, even though his four voyages had developed persuasive evidence that he had discovered completely new territories not part of the Indies, Japan, or Asia.

Nevertheless, Columbus was a giant of exploration, and he must be credited with "discovering" an area of the globe unknown to European or Asian civilization. His inaccurate hypothesis propelled him not to Asia but to rich islands and Central and South America. The careful preparations he had made for proposing his "enterprise of the Indies" were vital to the legitimate discovery of a new world he had never envisioned. Thoughtful observation allowed him to exploit the trade winds that blew him to America and back, and his peerless navigational skills allowed him to blaze the trail to harbors in America for himself and others.

Although he was born in 1451, Columbus was in many ways a very modern man. Raised in the cosmopolitan city state of Genoa by a middle-class family in the weaving and wine-purveying trades, Columbus went to sea at an early age. Records of his early life are sparse, but his first coastal voyage was in his early teens, and there are indications that he participated in both merchant expeditions and sea battles throughout the Mediterranean.

By his early 20s, Columbus was developing the personal qualities that made him such a notable achiever: a strong will, deep religious faith, courage, and burning ambition. He was acquiring the mastery of seamanship necessary to execute a voyage of discovery, and such ideas may have been already taking root in his mind. However, there was really no way voyages of discovery could be supported

How the Great Scientists Reasoned. DOI: http://dx.doi.org/10.1016/B978-0-12-398498-2.00003-5

by the declining, Mediterranean-centered Genoa, which was rapidly being stripped of her eastern trade empire by a surging Turkish Ottoman power.

Somehow, Columbus journeyed from beleaguered Genoa to Portugal, the world center of scientific voyages of exploration. Although disputed by some historians, the most vivid account of his arrival in Portugal is given by both Morison and Dor-Ner [1,2]. In 1476, Columbus sailed with a convoy carrying resinous mastic from Genoa's eastern outpost of Chios to Lisbon and the northern European ports beyond. Although the merchant ships were armed with cannons and were guarded by a ship of war, they were attacked by a pirate fleet on the Atlantic side of the Straits of Gibraltar. Defeated after a fierce daylong sea battle, Columbus's flaming ship sank, leaving the wounded Genoese to make his way to shore by grasping a floating sweep oar. He was nursed back to health by the kind villagers of Lagos, Portugal, and he then made his way to Lisbon.

Columbus found a new and secure base in the Genoese expatriate community in Lisbon, where his younger brother Bartholomew had established a mapmaking business. This indicates that fascination with geography and the incomplete knowledge of what lay to the west of Europe were family traits. There was a thriving demand for nautical charts in Portugal, and the brothers' business apparently did well. The few examples of Columbus's surviving charts show him as uncommonly skilled at this craft, and his early years of drawing maps of the world must have focused his mind on the limitations of fifteenth-century geographic knowledge. His sailing experience would teach him that the broad Atlantic, the "Ocean Sea," had many islands; his mapmaking taught him that geographic knowledge was limited and fallible. A professional chart maker would quickly realize that some areas of the maps of the time were little better than guesses.

Lisbon was Columbus's base from 1476 to 1485. Although he occupied himself largely with the mapmaking business, we can presume that he took any opportunity to expand his knowledge of the Ocean Sea, which the uneducated presumed to extend to infinity. Despite the death of Prince Henry the Navigator in 1460, discovery was in the air in late fifteenth-century Portugal as the great national effort to reach the Indies by circumnavigating Africa continued. A series of doughty mariners pushed the envelope of the known world farther and farther south down the coast of western Africa. Figure 3.1 shows the Azores, Canaries, and Cape Verde islands, which were occupied by Portugal in the fifteenth century; these islands were to be the staging areas for Columbus's voyages of discovery.

Breathing this bracing air of discovery, Columbus was not to be bound to a desk. Early in 1477, he gained a berth from Lisbon on a voyage "a hundred leagues beyond the island of Thule (Iceland)" of which we know little, but such a voyage would have been an ideal opportunity to absorb knowledge of the abandoned colony of Greenland and hear rumors of Vineland. The Norse expeditions to Vineland had probably ended in the twelfth century when global cooling made the colony less attractive. Greenland retained a livable temperature for many centuries afterward, but its pastures gradually grew less productive and more ice covered, with the colony totally abandoned only in the early fifteenth century. It is easy to imagine that some Icelandic seaman's knowledge of great lands to the west endured for

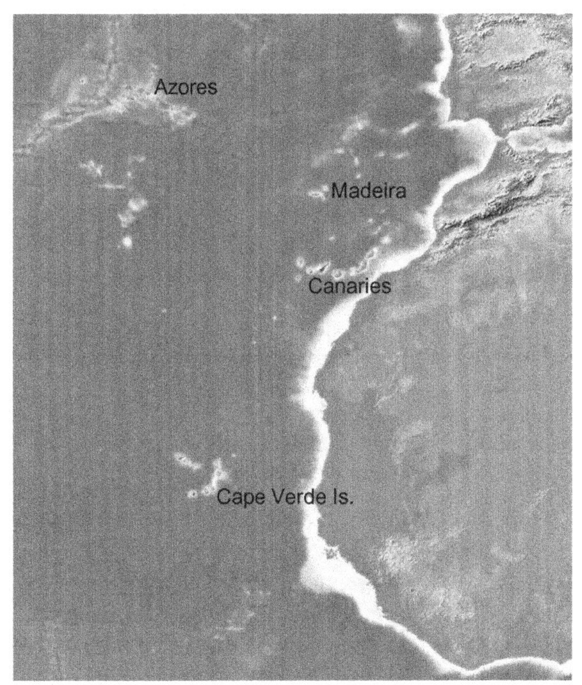

Figure 3.1 The Azores, Madeira, Canaries, and Cape Verde Islands. These became the staging areas for Columbus's explorations.

the half century required to bring it to Columbus's ears. After all, *The Saga of Eric the Red* was compiled by Jón Thórdharson [3] in about 1387 and has survived to the present day. Moreover, in Iceland, Columbus would have made direct contact with the same English and Irish Icelandic trade that inspired the discoveries of John Cabot.

As Columbus waxed intellectually in this invigorating climate of Lisbon, he learned Portuguese, Castilian, and Latin. We know of one merchant voyage to Madeira and Genoa that Columbus captained in 1478. Somehow, the poor immigrant youth established himself well enough in Lisbon society to merit a marriage to the wellborn daughter of an aristocratic Lisbon family, Dona Felipa Perestrello e Moniz. Columbus was evidently learning to be persuasive, an important prerequisite for a career that would require him to convince sovereigns that his expeditions would bring wealth to their kingdoms.

His marriage gave Columbus several key advantages that were crucial to his later voyages. For one thing, Felipa gave Columbus his only legitimate son, Diego, born about 1480. Dona Felipa's father, Perestrello, who died before the marriage, had been a skilled seaman, educated at Prince Henry's navigation school. Perestrello had taken part in the second colonizing expedition to the Madeira archipelago (Figure 3.1) and had been rewarded with the governorship of Porto Santo, the smaller of the two major islands. The young couple sailed to Madeira immediately after the wedding and spent 2 years there, ostensibly to give Columbus the opportunity to develop an income stream for Felipa's needy mother and brother.

Columbus's interest in trade was probably much less than his fascination with Perestrello's abundant charts, logs, and records of Atlantic voyages. They were fine fodder for Columbus's vivid imagination.

During these quiet years in Madeira, Columbus must have had abundant time to look out over the Ocean Sea as he assimilated Perestrello's papers and dreamed about what further lands might be found there. Clues about strange territories might even be gleaned from the flotsam washing up after westerly storms on beaches in Madeira and the Azores. Columbus collected the following list from his fellow islanders that he later used to buttress his arguments for the western route to India [1, p. 60]:

- "A piece of wood, ingeniously wrought, but not with iron" collected in the Azores.
- A similar piece collected at Porto Santo.
- Canes larger in diameter than known to grow in Africa.
- Strange seed pods now known to be a horse bean that originated in Central America.
- Two dead bodies resembling "Chinese" washed ashore in the Azores.

What could all this signify if not that Asia lay tantalizingly close across the Atlantic?

Furthermore, Columbus made two key observations that had been part of Portuguese navigational lore and were fundamental to his successful voyages. He noted how steady were the westerly winds that blew at the 33°N latitude of Madiera, month after month. Yet in the Canaries at 28°N, the prevailing trade winds were from the northeast. Columbus made the reasonable inference that the prevailing easterlies could endure far enough for a westward sail to the Indies and that the westerlies that could be accessed at higher latitudes could propel a return journey. It would be difficult to overestimate the importance of this discovery because these winds were to be the European path to Caribbean discovery and the exploration of the New World.

The prevailing wind directions were important because the square-rigged caravels of the late fifteenth century sailed best with a wind at their back; their square sails were fixed to braces having restricted play perpendicular to the ship's keel. Attempting to sail into the wind's direction was impossible, and heading into the wind was accomplished only with lengthy tacking. A second option was a lateen-rigged ship, which has triangular sails attached to a boom that can orient itself freely. Such rigging can beat much closer to the wind should tacking become necessary. However, in sailing with the wind dead astern, a lateen-rigged ship is subject to the shock of jibing because a small variation in the direction of a following wind may destructively force both sail and boom to the opposite side of the ship. Columbus knew that even his most optimistic assessment of the distance to the Indies was long enough so that he could only make the crossing with a good tailwind. Calculations show that Columbus needed to have the wind nearly at his back for the whole trip if he were to make his calculated distance to the Indies in about a month.

Sometime after 1481, Columbus sailed to the newly built Portuguese fortress of Mina on the west African Gold Coast. Columbus's surviving notes in the margins of Aeneas Sylvius's *Historia Rerum* draw on these experiences to debunk the notion that these equatorial regions are uninhabitable, indicating his accumulating experience that the contents of books, even those written by influential popes, are sometimes wrong [1, p. 41]. Columbus had learned another fundamental rule of scientific thinking: Your own careful observations are more likely to be true than the published opinions of armchair experts.

And as the long reach of Africa into the southern hemisphere became better documented, the eastern route to the Indies may have looked increasingly less practical to Columbus. During the early 1480s, Columbus continued to develop his hypothesis that ships could sail west from Europe to access the trade riches of India while avoiding the far longer easterly route that was under active development. The idea of a spherical earth and consequent existence of a western route to the Indies was not novel; in fact, several geographers of the time had discussed it [2, p. 72]. We know that Columbus exchanged letters with the Florentine intellectual Paolo dal Pozzo Toscanelli, who accepted the claims of Marco Polo for a long Eurasia, implying that the Indies might be reached by a short western voyage. However, Columbus and Toscanelli differed from conventional thinking because they were convinced that the western ocean between Europe and Japan was so limited in extent that crossing it was feasible with the caravels of that period.

The major competing notion was based on the recently rediscovered Geographia of Ptolemy [4]. This second-century Alexandrine Greek had compiled the knowledge of the ancient world to argue that the ratio of land to sea should be about one-sixth (not a bad guess!), a belief much more compatible with a wide western ocean between Europe and Asia. In response, as "the enterprise of the Indies" developed in Columbus's mind, he liked to cite his own biblical evidence from the book of Esdras (included in the Vulgate but not the King James version of the Bible) that the world is mostly land. Esdras's retelling of Creation Week includes the passage, "And on the third day Ye united the waters and the earth's seventh part, and dried the six other parts." So the Bible tells us that earth's surface is only one-seventh water. It is a telling coincidence that these two "authorities" came up with numbers that are almost reciprocals of each other!

So how far west from the Iberian Peninsula did Columbus expect to sail to reach India? The circumference of earth had been determined to good accuracy by the Greek Philosopher Eratosthenes in about 200 BC. He compared the noon altitude of the sun at Alexandria and Aswan at the summer solstice (Figure 3.2); knowing the distance between these locations and assuming Alexandria lay due north of Aswan, he calculated a value not far from our currently accepted 25,000 US miles. Many other estimates had been made since, but Columbus selected one of the smallest: that of the medieval Islamic geographer Alfragan [1, p. 65]. Columbus compounded his optimism by converting Alfragan's value incorrectly using the short Italian mile (which is only 91% as long as a standard mile). These two assumptions led Columbus to calculate a circumference of earth of only 75% of its true value. Since Columbus intended to sail due west from the Canary Islands at 28°N latitude, he understood that he could legitimately expect his journey to be 12% shorter than if he were sailing along a great circle route like the equator.

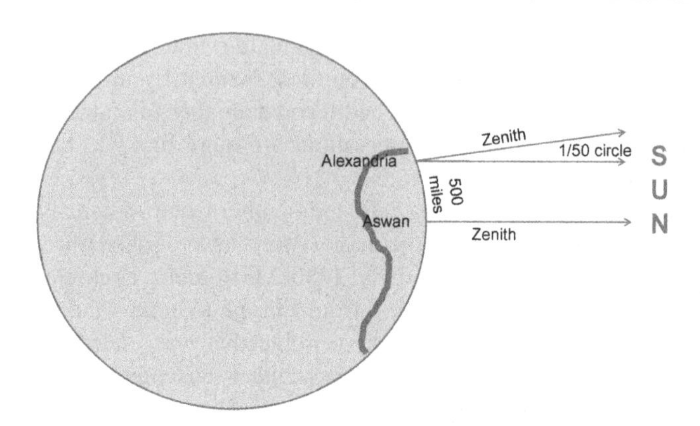

Figure 3.2 Eratosthenes' observation that moving due north from Aswan to Alexandria increased the angle from the sun to the zenith by 1/50 of a circle. The calculation corresponds to moving 1/50 of the circumference of the globe. Therefore, the circumference of earth should be 50 times the distance between the two cities, which Eratosthenes measured to be 500 miles.

The final and weakest link of Columbus's calculation was to subtract from earth's circumference the distance between the Iberian coast and Japan across Eurasia. He elected to use the most current data available to him: the records of Marco Polo. The tireless and excitable self-promoter Polo had considerably exaggerated the breadth of his travels to China, and Columbus augmented that distance for good measure. Moreover, Columbus defined his goal as "Cipangu" (Japan), and he somewhat overestimated its easterly separation from the Chinese mainland at 1200 miles instead of a true value of about 900 miles.

So Columbus used every tool he could: careful selection of the data, exaggeration of the selected data, and outright confused unit changes to make the distance from his projected starting point in the Canaries to Japan look manageable. Columbus's final estimate was 2400 nautical miles (2800 US miles), a gross underestimate of the true distance of 12,000 nautical miles. (We will use nautical miles throughout the following discussion.) If his caravels could advance at a steady 4 knots, he calculated that the enterprise of the Indies would be feasible. Imagine the consequences of depending on this specious calculation: Suppose no American continents barred the way as Columbus set out on his voyage to Japan. Columbus would have started from the Canaries at 18°W longitude and sailed due west past the current Bahamas after a voyage of 2400 miles. He would blithely sail through nonexistent Central America and begin his crossing of the immense space now occupied by the Pacific Ocean. After a brushing near the northern reaches of Taiwan (and probably missing that island altogether) far below the southern coast of Japan, Columbus's flotilla would arrive on the western coast of China after an epic voyage of 12,000 miles. Columbus's planned 2400-mile journey might be made in 25 days if fortuitous winds gave him an average speed of 4 knots, and the most optimistic case might be his own fourth voyage, which reached the Caribbean in only 20 days. Even with the assumption of such favorable winds, the 12,000 nautical mile distance would require an unfeasible journey of 125 days! Before his first voyage, Columbus instructed his quartermasters to prepare provisions for a voyage of 28 days, and any 125-day voyage would far exceed the water, wood, and food capacity of the tiny caravels (Figure 3.3).

To fund the enterprise of the Indies, Columbus exploited the connections of his in-laws, the Monizes, and the latent interest of the Portuguese monarchy in the western route to the Indies. The Monizes seem to have been blood relations with Fernando Martins, a dean of the Lisbon cathedral, who had nurtured a long-term fascination in the concept of a western route to the Indies. Martins had been to Italy, and his interest in cosmology and geography had also been stimulated by his acquaintance with Paolo Toscanelli, who had written a letter to the Portuguese crown describing a western route to the Indies some 10 years ago.

Fortified with his calculation and encouraged by the conclusions of Toscanelli, in 1484 Columbus presented a proposal to the shrewd and sophisticated Portuguese monarch Dom João II to take an expedition of more than one ship to discover a western route to the Indies. The terms he requested for making this voyage have not been preserved, but if they were as expensive as Columbus later demanded from Spain's Ferdinand and Isabella, they would have been extremely generous to Columbus. Dom João dealt with the problem as a government leader would today: He referred it to a commission of experts. He had only recently established a Junta

Figure 3.3 2005 Reconstruction of the *Santa Maria*.
Source: Photo by Dietrich Bartel.

dos Mathemáticos whose members included the cosmographer and bishop of Ceuta, Diogo Ortiz, and two other learned geographers.

After the preliminary evaluation, we may presume that the commission attacked Columbus's mathematical inconsistencies, his dependence on the estimates of the breadth of Asia from Marco Polo, and the fact that his destination of Cypango was not even known to Ptolemy. They reported to the king that Columbus's proposal was "vain, simply founded on imagination" [1, p. 71]. The news that the king had rejected the proposal came to Columbus within a year. What a delicious irony! The rejection by the Ortiz commission was much more scientifically accurate than Columbus was. With the exception of the existence of Japan, Columbus was profoundly wrong, yet the Portuguese failure to fund this expedition was a gross lapse of imagination that constricted the future of a great nation.

Perhaps part of the rejection stemmed from a competitive proposal. Columbus seems to have been underbid by a less-ambitious and expensive western exploration scheme by Fernão Dulmo, who wished to sail two caravels west from the Azores to discover the rumored isle of Antillia. Columbus knew enough about the winds that prevail at this higher latitude to understand that Dulmo would exhaust himself beating against the prevailing westerlies before he would discover anything. The outcome of Dulmo's voyage has been lost to history.

By late 1485, Columbus had assimilated this definite "no" from Dom João. He had also suffered through the death of his young wife, and his best efforts to generate family connections to marshal support for the enterprise of the Indies had met with final rejection. Picking himself up after these losses and showing a clear grasp of the geopolitical situation, Columbus left Portugal for Castile.

Ferdinand and Isabella, monarchs of the united kingdoms of Aragon and Castile, were the most likely rivals of Dom João. They were young, aggressive, and outward looking. Unfortunately, they were otherwise occupied, and their project was arguably as significant as Columbus's own. These two monarchs were marshaling all the military power of their newly unified kingdom to conquer the Moorish kingdom of Granada and successfully repel an invasion that had begun in the eighth century when the forces of Islam had quickly subjugated the Iberian Peninsula in their thrust toward the heart of Europe. Columbus was to spend seven years presenting and arguing his plan to the two monarchs and their scientific advisers, but the final funding decision was always kicked down the road until ultimate victory was achieved with the fall of the Moorish capital in early 1492.

Columbus's initial efforts to find Spanish support in the spring of 1486 were encouraging: He was received in the Alcazar of Cordova by Ferdinand and Isabella, who graciously listened to his exposition. A decision, of course, would depend on the crown's comptroller of finances and evaluation by a panel of learned experts chaired by Prior Hernando Talavera. After the commission took up its deliberations, Columbus was paid a small retainer from the crown, about the wages of a common seaman, during the following 2 years, but was left to dangle unsupported for 4 more years until it rendered its final verdict.

With the establishment of the Talavara commission, Columbus found himself with two problems, each very difficult to surmount. The first was that Talavera himself had a clear grasp of the best geographic knowledge of the time. This included a realistic assessment of earth's circumference and the breadth of Eurasia, and the difference between these numbers. Talavera knew that the distance a westward voyage to the Indies must cover was much closer to 12,000 miles than Columbus's estimate of 2500 miles.

The second was the heavy clerical representation on the committee. St. Augustine had already opined during the fourth century that earth was mostly covered by water. Ask yourself, with Torquemada achieving prominence in Spain and the inquisition gaining importance year by year, would you as a foreigner and speaking a rather primitive kind of Castilian have the temerity to contradict the authority of a revered saint on the extent of God's creation?

During these hearings before the commission, Columbus argued tenaciously for the geographical aspects of his project. Could he allow arithmetic based on imperfectly known parameters to defeat this burning ambition? But when attacked on theological grounds, besides trotting out the book of Esdras, he was nearly tongue tied. Perhaps he was having visions of the inquisition and how easily a slip of his imperfect Castilian could focus Torquemada's attention on his potential heresy.

During the on-again, off-again years of negotiation, Columbus did briefly return to Lisbon to witness the celebration of the return of Bartholomew Dias's fleet in December 1488 after rounding the Cape of Good Hope to briefly sail into the Indian Ocean. How the fires of his ambition must have burned even more fiercely on seeing Dias's elevation. This success certainly reaffirmed Dom João's commitment to the eastern route and must have underscored the futility of seeking support in Portugal for a western trade route. His future must lay in Castile.

The Talavera commission probably rendered its final verdict in 1490 in Seville. Columbus's sixteenth-century biographer Las Casas reported that they "judged his promises and offers vain and worthy of rejection" and the project "appeared impossible to any educated person, however little learning he might have." Ouch, that was meant to be a stinging rebuke!

The following were among specific points cited [1, pp. 97–98]:

- A voyage to Asia would require 3 years.
- The western ocean is infinite and unnavigable.
- St. Augustine teaches that the greater part of the globe is water.
- So many centuries after the creation, it is unlikely that anyone could find valuable new lands.
- If one were to reach the Antipodes, it would be impossible to return.

The final point about the Antipodes relates to the medieval notion that on the diametrically opposite side of the globe something weird has to happen to keep you from falling "down." Some authors speculated that Antipodal people's heads may be touching the ground with their legs in the air. Once you make this polarity switch, you may never get back!

Even after receiving this damning report, Ferdinand and Isabella did not make a final decision; they just informed Columbus that judgment would be deferred until the conclusion of the war.

Imagine Columbus's frustration, gradually turning to seething anger. He had just wasted 5 more years of his life waiting for a commission report, and those unimaginative royal fops still were unable to make a decision. Columbus spent the next 2 years on tenterhooks. The siege of Granada dragged on, looked encouraging, and yet seemed interminable. In desperation, frustrated Columbus made overtures to France and England through the offices of his brother Bartholomew. Doubtless his heart sank at the prospect of beginning the lengthy search for royal support again in a third country, but he simply could not squander more of his life in waiting.

In this moment of extremity, Columbus's luck changed. In late 1491, negotiations were beginning for the surrender of Granada. Moreover, the royal treasurer was imaginative enough to argue to Isabella that what Columbus may gain for the crown might dwarf the costs of the proposed expedition. Columbus was summoned to the court sitting outside Granada.

At this moment, Columbus felt he had waited and suffered enough to risk driving a hard bargain with Ferdinand and Isabella. They gulped at his demands, feigned refusal, and finally accepted the terms. We are fortunate enough to have surviving copy of the agreement between the sovereigns and Columbus (cited by his Spanish name of Don Cristóbal Colón) [2, pp. 103–107]. Among other provisions, it states that:

- the sovereigns would provide and equip three caravels,
- Colón and his heirs were to be appointed admirals over his discoveries,
- Colón was to be appointed viceroy and governor general over the new lands
- Colón would keep 1/10 of the gold, silver, and products of these domains.

Of course the sovereigns could not really keep an agreement so lucrative to a commoner and a foreigner, and this document was to become the basis of a lawsuit between Columbus's heirs and the crown that continued for decades after his death.

Provided that the expedition would be successful, Columbus would finally vault his family above its humble origins in weaving and wine retailing and into the aristocracy with a hereditary title for himself and growing riches for his family. Ferdinand and Isabella, having decided to provide the three ships, instructed that the expedition was to depart sooner than was really feasible. Columbus's plausible calculation had justified a voyage; now he had the resources to execute the plan.

Columbus was directed to proceed with all possible speed to the small seaport of Palos to outfit three caravels using the crown's resources. In Palos, enthusiasm for the project was sufficient to gain him support from Martín Pinzon, an influential local captain, and to recruit some first-class sailors hoping to share in the riches of the Indies. After refitting a lateen-rigged caravel to square rigging at his only stop in the Canaries, Columbus set out through the trackless miles of a virginally empty Atlantic, with plans to sail due west at 28° latitude until he reached Japan.

Using the astrolabe or quadrant should have allowed Columbus to determine and hold his latitude, but his skill in celestial navigation seems to have been rather limited, and his journals document several grossly inaccurate sightings of Polaris. It seems that his major problem was that he frequently did not recognize the pole star. His inaccuracies would have been compounded by the fact that in the fifteenth century Polaris was positioned more than 3° away from the polar axis of the earth, whereas the relative motion of earth and Polaris have positioned it now to within 1°. Figure 3.4 shows that, in concept, it should be fairly easy to approximately determine latitude without complex instruments by measuring the angular distance (θ) of Polaris above the northern horizon. However, Columbus was a better naked-eye celestial navigator than he was at using his astrolabe. He noticed very correctly on his return journey as he approached Lisbon harbor that Polaris seemed to be at the same altitude as he had previously observed while residing in Portugal.

Fortunately, Columbus ignored these frequently specious celestial observations and adhered to the results of dead-reckoning navigation, where his skills were superb. In this era, before the wide adaptation of the *log* that was cast from the moving ship, mariners apparently guessed at the average speed they were making through the water, perhaps adding a correction for any sizable ocean currents and multiplying by the duration of their daily run. The direction of this daily distance vector was determined by the ship's magnetic compass. For Atlantic crossings, the compass corrections required for the difference between earth's magnetic and geographic north poles are tolerable compared to the errors endemic to the above methods. The navigator simply laid out this vector on the chart with dividers to update his daily position. As you might imagine, course and velocity changes could make this procedure fairly tedious, but Columbus's mastery gave him quite acceptable results.

We marvel now at the accuracy of his "guessed" velocity, but perhaps he used telling clues, such as characteristics of the bow wave or the ship's wake, to determine his speed. Our advances in the hard knowledge of navigation may have speeded the loss of this "softer" and yet very valuable navigational knowledge. There are no clues to his methods in the surviving vestiges of his logbooks, and the originals he submitted to King Ferdinand were lost in the cataclysm of the Napoleonic wars.

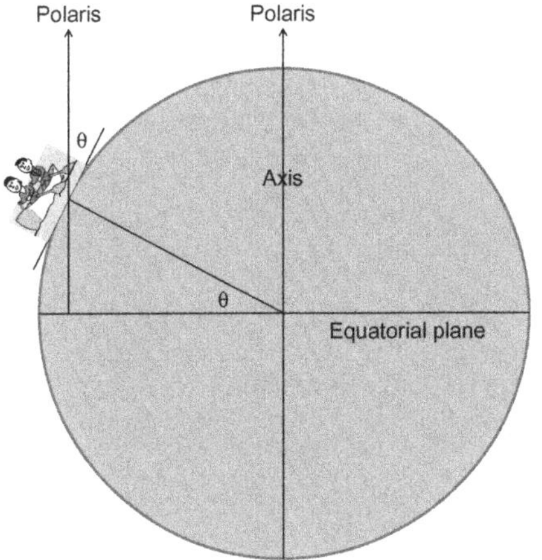

Figure 3.4 Latitude of the boat, θ. The latitude is determined to be the angle between the horizon and Polaris, or the complement of the angle between the zenith and Polaris.

The years of poverty and rejection Columbus endured certainly did strengthen his resolve not to turn back until his goal was achieved. After the 30th outbound day, a conference was held: The worried crewmen favored returning, but the captains gave limited support for Columbus. As he argued to continue for four more days, Columbus must have been mindful of the succession of Portuguese captains that Prince Henry patiently sent down the west coast of Africa—all returning when the terrors of the unknown became overpowering. He had invested too many years of struggle to allow his fleet to meet with failure. By the 33rd outbound day, evidence of land birds and vegetation encouraged the sailors enough so that opposition was overwhelmed by anticipation.

The landfall in the Bahamas on October 12, 1492, on the 37th outbound day, was thus the culmination of more than a decade's work for Columbus. The initial meeting with the native Arawak-speaking "Indians" on the tiny island of San Salvador was peaceful enough to bode well for the future. Imagine Columbus's elation; he had found the outlying islands of the Indies almost exactly where he had calculated. Sovereigns had ignored him. Professors had spurned him. Courtiers had mocked him. The experts of two kingdoms were wrong, and he alone was right. Even a less-egotistical man than Columbus could have had his head turned by a success of this magnitude; it could be a lifetime high for any man.

However, Columbus was far from the finish line. As he continued his voyage to islands farther to the south (Figure 3.5) and ultimately to the northern coast of Cuba, the pressure started to build. Where was Japan? Where were the temple roofs of precious metals? If they were in China, where was the Great Kahn? Most important, where was the gold? Columbus's fertile imagination dealt with these problems

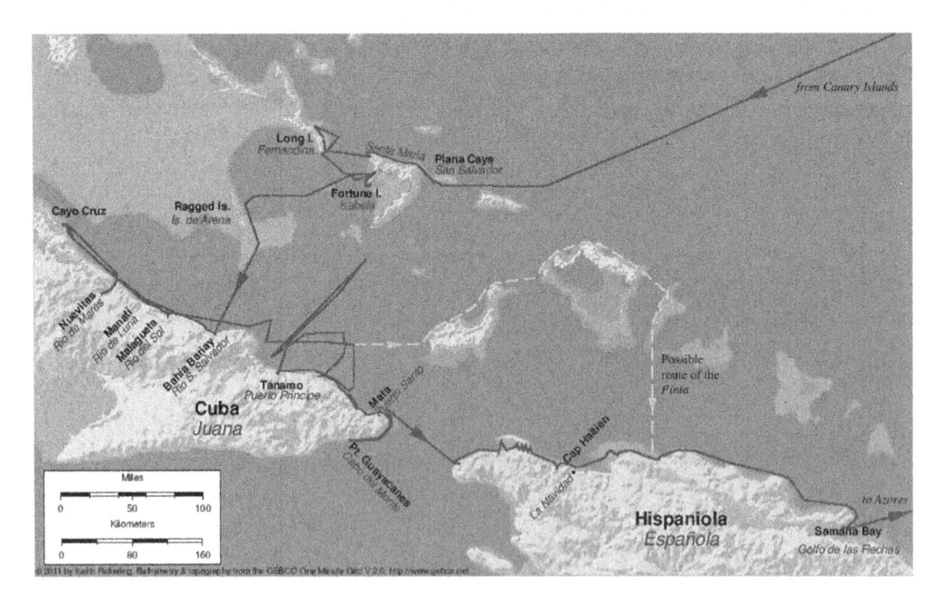

Figure 3.5 Columbus's first voyage.
Source: Keith Pickering and GEBCO.

by assigning the Arawak word *Cibao* (Cuba) to *Chipangu* and their word *Nancan* (village) to *Kahn*, continually raising his hopes that a meeting with representatives of the eastern empires was imminent. Meanwhile, he traded vigorously, and later ruthlessly, to acquire every speck of gold he saw. But the mania for gold waxed and spread uncontrollably to the men. When the three caravels crossed the Windward Passage and began exploring the north coast of Hispaniola for the abundance of gold promised by the Arawaks, the lust for gold became a flame that threatened the expedition with disaster.

Captain Pinzon, on the pretext that rough weather separated him from the *Santa Maria* and the *Nina*, broke contact with the other two ships to sail his faster caravel *Pinta* ahead to investigate rumors of more gold. Columbus first waited for him to return to rendezvous with the other two craft, then accumulated more encouraging stories of gold from the inhabitants, and finally forged ahead in too great a hurry to overtake Pinzon. Sailing into the teeth of the prevailing westerlies, Columbus was forced to take advantage of the evening land breezes to make easterly headway, necessitating hazardous passages through jagged rocks immediately off the north shore of Hispaniola. Late on Christmas Eve of 1492, the luck of the *Santa Maria* finally ran out as she lodged on jagged rocks; by dawn she was rapidly breaking apart. With the aid of the natives, the crew salvaged what they could from the wreck and built a fort on high ground nearby. Columbus established good relations with the local caique, recruited volunteers to remain (and search for gold), and left this crew to await a rescue voyage. Luckily, the *Pinta* returned shortly thereafter; the necessity of provisioning the castaways at their new village of Navidad with as much of the remaining food as possible dictated an immediate return of the remaining two caravels to Spain.

In January 1493, the *Niña* and *Pinta* sailed from Hispaniola on a northeasterly tack, hoping to leave the zone of the easterly trades and find westerly winds. The return voyage took place near the latitude of the Azores, about 38°N. At first the winds were hopefully brisk and favorably directed, but as the two caravels approached the Azores, they ran into hurricane-force gales that drenched the crews continuously with cold seawater. Needing urgent relief near the end of this storm-tossed return, Columbus calculated the position of the *Niña* to be 75 miles south of the Azores—off by only 25 miles. By comparison, the independent observations of the *Niña*'s captain Vincente Pinzón, the amateur astronomer Bartolomé Roldán, and apprentice pilot Perlonso Niño were each hundreds of miles off, though in different directions.

So even though navigation was a less-critical skill on Columbus's outbound first voyage, it was crucial to his return to Spain. Despite being buffeted by a series of hurricanes during his winter crossing of the North Atlantic, Columbus was able to find the Azores to gain respite from the storms and later pilot his way into the stony Tagus entry to Lisbon harbor with a ship having only one remaining scrap of sail.

After Columbus's return to Spain, crowds thronged his procession to the court at Barcelona; Ferdinand and Isabella generously feted him as a hero, allowing him even to be seated in the royal presence. And what a sensation Columbus was, wowing both commoners and courtiers with his exotic Indians, colorful parrots, and artfully displayed gold jewelry. The sovereigns were so pleased that they authorized a second voyage to relieve the castaways and begin colonization of Hispaniola. Columbus enjoyed the high point of his career as he reported to their majesties that he had discovered the Malay Peninsula (Cuba), which was known to be connected to China, and the wealthy island of Japan (Hispaniola). Perhaps, he admitted that a little future exploration needed to be done to certify the details, but this was the general picture.

The meteoric success of this foreign-born commoner naturally aroused much envy at the court. In fact, his rather notable rapport with Isabella rankled the king. One thing Ferdinand—by the grace of God, king of Aragon and defender of the faith—did not want to hear from his beautiful queen across the breakfast table was rhapsodizing about the courageous Italian admiral.

But Columbus was off too rapidly to see the storm clouds gathering. In late September 1493, Columbus sailed with an armada of 27 ships carrying 1200 volunteers, mostly lured by lust for gold. It was typical of Columbus's mania that he exploited this opportunity to claim some new islands before relieving his marooned garrison or landing his settlers. Therefore, he set his course somewhat to the south of Hispaniola and made landfall in the Leeward Islands just south of Guadalupe, shown in Figure 3.6. Columbus had heard from the natives on Hispaniola of the beauty and wealth of these territories, and he aimed to burnish his accomplishments by acquiring them for the Spanish crown. Exploring his way northwest through the Virgin Islands, Puerto Rico, and the northern shore of Hispaniola, Columbus finally arrived at his settlement of Navidad in November 1493 to find that the entire

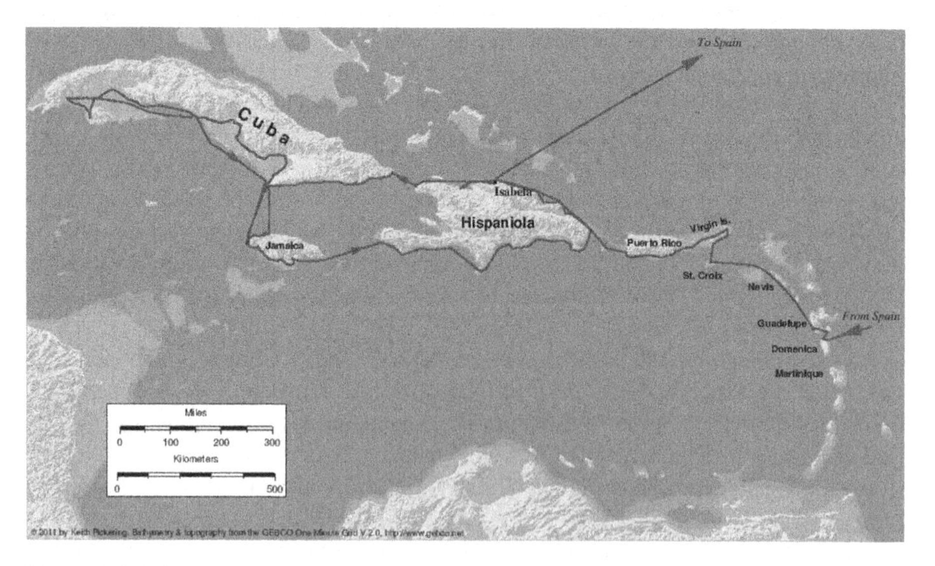

Figure 3.6 Columbus's second voyage.
Source: Keith Pickering and GEBCO.

garrison of 40 men had been wiped out. Columbus interrogated the nearby natives, receiving a conflicting set of stories casting the blame on a Carib raid, but the real reason the natives had turned on the Spanish was not hard to discern: conflict over food, gold, and women. Again, the resolution of this situation was pure Columbus: resettle his colonists to a precipitately chosen and similarly fever-ridden site nearby that he called *Isabela* and set sail with three caravels to discover new territory. Whatever Columbus had told the Catholic monarchs, he himself evidently needed further proof.

After crossing the Windward Passage, Columbus elected to sail west along the southern, previously unexplored, coast of Cuba. Since he believed that the eastern tip of Cuba was the western extremity of Asia, probably the Malay Peninsula, if he could demonstrate that Cuba was not an island, then he could finally resolve lingering doubts that he had reached the (East) Indies. At first the plan went well. The coast was beautiful with good anchorages. Columbus, however, interrupted his coasting for his top priority; a weeklong detour to Jamaica on the strength of the natives' assertion that much gold was to found. Once again he found the natives' urge to oblige him greater than their veracity, and he found little gold in Jamaica.

Resuming his coasting of Cuba, Columbus found that navigation became progressively more difficult, with perilous reefs, numerous small islands, and stormy weather. These conditions terrified the crew, because mariners' legends had prophesied that at its outer limits the Ocean Sea shaded into unending and dangerous shallows. Supplies began running low, and the crew began murmuring. A Cuban native told him that the coast continued so far that one could not reach its end for 40 months.

These facts had boxed Columbus in; he must return to Isabela. Ironically, the point where he turned around was only about 100 miles from the western extremity of Cuba. Could it be that he had heard a rumor from some other, more truthful Cuban that a hypothesis-busting discovery lay just beyond the next cape? We will never know, but turning around so close to gaining significant proof that 10 years' worth of assertions and arguments were incorrect would be a great coincidence. Nevertheless, Columbus's official view at this time that this voyage reaffirmed that he had sailed up the Malay Peninsula almost to China.

Before Columbus turned his ships back toward Hispaniola, he followed the example of Bartholomew Dias, who had exculpated himself before returning to Portugal shortly after rounding the Cape of Good Hope. Columbus sent his fleet notary to all the professionals and seamen requiring them to attest "if they had any doubts as to whether this land was the continent of the Indies." The oaths were all sworn; no man in the fleet doubted that they had reached the Indies. So much for "proving" facts by affidavit.

Perhaps it was a sign of Columbus's inner conflict over these issues that he fell seriously ill on the return voyage. He became incoherent, fell into a coma, and was incapacitated by gout and rheumatism once awake. He suffered periods of serious disability for the rest of his life.

His return to Hispaniola was a further disappointment. Incessant rain and flooding had spoiled many of the provisions, so the colonists were forced to subsist on unappetizing manioc tortillas. It was becoming unlikely that a major source of gold would be found near the Isabela settlement. Columbus worked with his brother Bartholomew to bring discipline and order, but neither of them were able leaders, and they were more apt to hang troublemakers than inspire their cooperation. The natives were pressed so hard for food and gold that a small-scale revolt spun into open warfare. By this time, Columbus had left a long trail of disaffected captains and colonists, some of whom had returned to Spain and pleaded their case. The sovereigns sent a royal commissioner to straighten things out; Columbus adamantly declined to cooperate and returned to Spain in 1496.

Ferdinand and Isabella knew they had a problem on their hands as long as Columbus governed because it was becoming abundantly clear that he had no temperament for leadership or administration. Nevertheless, after allowing him to stew for 4 months after his return, they consented to see him and treated him with much of the usual courtesy. Columbus pleaded for more colonists, ships, and supplies. The sovereigns replied that they were supporting 500 men in Hispaniola at great cost already. Luckily for Columbus, the sovereigns had to adapt to the Treaty of Tordesillas, which created a demarcation line in the Atlantic between the newly discovered territories belonging to Portugal and Spain. Lands lying east of this line would belong to Portugal, and lands lying to the west would belong to Spain. Columbus was the obvious candidate to look for a large landmass in the southern seas that might lie on the Portuguese side of this demarcation line and threaten Spanish interests. He was assigned three caravels to investigate the equatorial region west of the Cape Verde Islands, which lie west of Africa's westernmost bulge (Figure 3.1); this was to become Columbus's third voyage to the Caribbean.

This exploration began in May 1498 with Columbus in high dudgeon. He remarked in his log, "the course I am following has been taken by no one else and these seas are entirely unknown." However, anticipation faded as tropical hardships developed, and the ships were becalmed in such blazing heat that the crew feared their ships would catch fire. The provisions spoiled, wine and water casks burst, and Columbus's maladies returned in disabling form.

In his extremity, Columbus headed his ships northwest just as he was approaching the Orinoco delta, now in Venezuela. The expedition's first landfall was the south coast of Trinidad (Figure 3.7), and as the ships sailed west into the Gulf of Paria it became clearer and clearer because of the quantity of fresh Orinoco water in the sea that they were near the estuary of a river so mighty that it must drain a continent. Columbus's febrile imagination now came to the mystic conclusion that these waters originated from the Garden of Eden, though the garden itself may have been at a considerable distance inland. As soon as he extricated himself from the "Dragon's Mouth" at the northern end of the gulf, he headed for Hispaniola.

He arrived in Bartholomew's new and excellently sited city of Santo Domingo in August 1498 to find the fires of rebellion burning with renewed vigor. Francisco Roldán, a charismatic former mayor of Isabela, had forged an alliance with the natives to resist the gold-seeking administration of Bartholomew Columbus. The Columbus brothers tried both the mailed-fist approach, which seemed to be leading to the depopulation of the colony, and the velvet-glove approach of accommodation

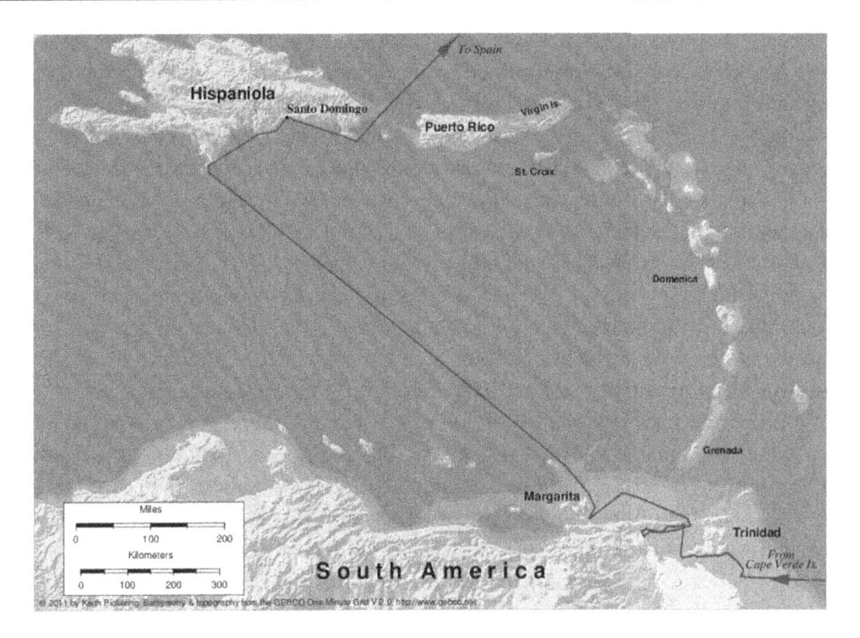

Figure 3.7 Columbus's third voyage.
Source: Keith Pickering and GEBCO.

but could make neither work. Their initial heavy-handed campaign to stamp out resistance had simply created too much disaffection. Columbus wrote a letter to Ferdinand and Isabella asking for permission to send the troublesome home, pleading for replacement colonists who had not served prison time and for "a wise administrator of justice." In response, the crown sent the influential Francisco de Bobadilla with powers to assume full authority over the colony. Columbus refused to cede power, cooperate, or compromise in any way, and he was returned to Spain in chains in October 1500.

After letting Columbus cool off for 6 months, Ferdinand and Isabella consented to see him and later somewhat mollified him. It was becoming clear to them that these new lands were stupendously rich, and consequently they could expend more patience on the admiral than they might otherwise lavish on a failed viceroy. Columbus spent the succeeding year meditating on his achievements and submitted a series of requests for the funding for a fourth voyage that would find a passage through the Indies to the Asian mainland.

By the time his fourth voyage was approved, Columbus's health had continued to fail. He probably realized this would be his last exploration, yet he had a clear vision of what he must accomplish to quell the growing doubts that his discoveries were not in the (East) Indies. He planned to sail south of the Malay Peninsula to reach the coast of China itself. In the actual world of the Caribbean, his plan was to sail to the west far enough below the south coast of Cuba to avoid those treacherous shallows. Columbus's three caravels thus approached the coast of Central America near the Honduran coast as shown in Figure 3.8. Arduous days of sailing south along this coast showed no evidence of Chinese civilization, and efforts to break though Central America to reach the Chinese coast were vain. Detouring once again to search for a rumored gold mine, Columbus's fleet entered a lagoon that offered shelter from the prevailing easterly winds—while harboring voracious shipworms that holed his ships' planking. A desperate effort to sail the two remaining ships back to Hispaniola proved too risky, and Columbus was forced to beach his leaking ships in Jamaica. Living there under precarious conditions, he sent courageous volunteers by canoe to seek succor from Hispaniola but was forced to wait for almost a year for the dilatory acting governor to send a relief expedition. An ailing Columbus finally reached Hispaniola only in the summer of 1504.

Columbus returned to Spain and lived his two remaining years in waning health but growing affluence due to his proceeds from the crown. In the years following his 1504 return to Spain, Columbus clung to the idea that he had discovered a "new world" constituting the outlying islands of Asia. As the fifteenth century waned, Ferdinand was already sending new expeditions to exploit the pearl fisheries of the Venezuelan coast that Columbus had discovered in his third voyage. Amerigo Vespucci was a crew member on one of these expeditions and may have captained a second Portuguese expedition that explored much of the coast of Brazil [2, pp. 300–305]. On his return, certain that Brazil constituted a new continent, Vespucci did a much better job of documenting and publicizing his experiences than Columbus had and widely disseminated an account of his voyages. Amerigo

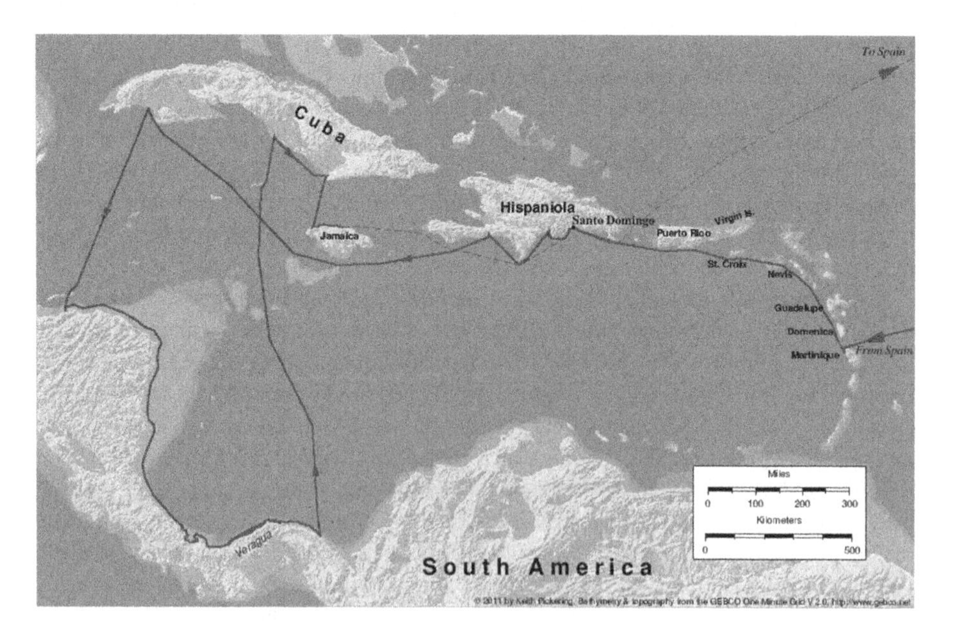

Figure 3.8 Columbus's fourth voyage.
Source: Keith Pickering and GEBCO.

Vespucci's name, for most of the next 200 years, supplanted Columbus's as the discoverer of the New World.

The life of Columbus should be an inspiration to discoverers, but it is rich with irony. Columbus conceived his voyages at the center of world sophistication for nautical exploration. But Portugal was *too* knowledgeable to accept Columbus's flawed calculation of the distance to the Indies, forcing him to seek funding in Spain. Spain might have rejected his proposal also had it not been for the burst of self-confident vigor released by the fall of Grenada. Turned down for many years by government commissions on both serious and fatuous grounds, Columbus finally prevailed and exploited his superior seamanship to discover South and Central America and the rich empire of the West Indies.

Columbus's conflict illuminates a problem endemic to applying the scientific method. Once a scientist commits himself to experimental study of a hypothesis, he generally has a stake in proving or disproving it. In the case of Columbus, he had invested years of mastering seamanship, navigation, mapmaking, and correspondence with the noted geographers of his day. He had dedicated his life to reaching the Indies and vaulting his family from its humble origins to membership in the titled hereditary aristocracy.

Columbus's struggles also illustrate the robust strength of the scientific method. Investigating even a false hypothesis can lead to precious new knowledge. In a very fundamental way, beginning even with a hypothesis very wide of the mark can still lead to valuable facts as long as the hypothesis can be updated to be consistent with new data as they become available. This might be compared to a method of successive approximations in mathematics: It converges rapidly to a correct solution if the initial guess is close to correct, but it may still approach an accurate answer even with a poor starting approximation, although it may take more iterations.

So we have seen that Columbus was courageous, brilliant, imaginative, and totally absorbed by and dedicated to discovery. He was a skeptical and independent thinker in an age when the authorities had license to painfully contort the bodies of those who dared to be too independent. These qualities would have made Columbus a fine scientist in any age.

But his yearning for success, his lust for advancement, or perhaps his own narcissism, made it impossible for him to accommodate changes to his ideas, so he clung the belief that he had visited the (East) Indies long after he should have.

In evaluating Columbus's achievements, we must remember that if he had performed a judicious evaluation of the most probable sailing distance to the Indies, he never would have made his momentous voyage. He was no pipe-smoking college Professor, although ironically he did introduce pipe smoking to Europe. He was a man of action and ambition, burning to make a place for himself and his descendants in the world of aristocracy and privilege that he had often visited only by sufferance and which had painfully rejected him for so many years. But he must have been driven by more than just personal ambition, as the sea had had a grip on him from his adolescent years. He wanted—in fact, needed—to know what lay beyond, and he had the courage to gamble his life to discover it. He was a giant.

References

[1] S.E. Morison, Admiral of the Ocean Sea: A life of Christopher Columbus, Little, Brown & Co., Boston, MA, 1942, pp. 23–24.
[2] Z. Dor-Ner, Columbus and the Age of Discovery, William Morrow & Co., New York, NY, 1991, p. 50.
[3] J. Thórdharson, The Saga of Eric the Red. In The Harvard Classics: American Historical Documents, P.F. Collier & Son, New York, NY, 1939, pp. 5–27.
[4] J. Ptolemy, Ptolemy's "Geography": An Annotated Translation of the Theoretical Chapters (Trans. L. Berggren and A. Jones) Princeton University Press, Princeton, NJ, 2000.

Bibliography

Z. Dor-Ner, Columbus and the Age of Discovery, William Morrow & Co., New York, NY, 1991.
Beautiful maps and graphics.
S.E. Morrison, Admiral of the Ocean Sea: A Life of Christopher Columbus, MJF Books, New York, NY, 1942.
Everything anyone would want to know about Columbus and fifteenth-century sailing customs and technology.
G. Granzotto, Christopher Columbus: The Dream and the Obsession, Doubleday, New York, NY, 1985.

4 Antoine Lavoisier and Joseph Priestley Both Test the Befuddling Phlogiston Theory: Junking a Confusing Hypothesis May Be Necessary to Clear the Way for New and Productive Science

This chapter will sketch the history of a really bad theory: the theory of phlogiston. This theory was not bad merely because the numbers were grossly diddled, as we saw in the chapter on Columbus, but bad because the underlying explanation was utterly and confusingly wrong and only by dragging it to the junk heap could modern chemistry emerge. This theory, a false step chemistry took as it climbed beyond its alchemical origins, was not totally discredited for 150 years.

Of course, we have to put this criticism in perspective: Understanding how many different elements comprise the chemical compounds and mixtures of which our world is made is a very knotty problem, and it is hardly surprising that it took hundreds of years to complete. In this chapter we will describe how Priestley, Lavoisier, and other chemists of the eighteenth century took the first giant step toward identifying the chemical elements.

In the Western world, the classical search for the fundamental elements began with the ancient Greeks, who speculated about the smallest divisions of matter. These ideas were codified by Plato [1], who proposed that "earth, air, fire and water" were the constituents of all matter. The idea of four elements is not very useful for understanding chemistry, but it pretty much nailed the four states of matter: solid, gas, plasma, and liquid (Plasmas are gases whose energies are so high that the atoms can decompose into charged ions and electrons.) This theory just does not help you very much if you wish to alter the *chemical* composition of matter, and real chemical knowledge had to wait for the late Middle Ages.

Chemistry was ripe for rapid development as technology advanced during the Renaissance since it is the most necessary discipline for many of humans' material needs. In seventeenth-century Europe, the precarious and competitive situation of the small German states led their princes to sponsor chemical research to try to stimulate their key industries: mining, cloth manufacturing, dyeing, ceramics, and

How the Great Scientists Reasoned. DOI: http://dx.doi.org/10.1016/B978-0-12-398498-2.00004-7

brewing. This led to rapid expansion of knowledge of practical chemistry just as enthusiasm for alchemy was waning.

As the early practical chemists became more sophisticated than the ancients, it became clear that there must be more than four fundamental elements. To German scientist George Stahl (1660–1734), it seemed reasonable that some element departed or was driven off during the process of burning [2, pp. 78–84]. Do not minimize that guess; it does look very consistent with the kind of everyday combustion that you observe on your hearth. It would not be hard to convince yourself when watching a burning log sheathed in flame that something is pouring out of the log. According to Stahl, *phlogiston*, coined from the Greek word for fire, is the constituent of all matter that is driven out during combustion, leaving just "calx" behind. Charcoal, in particular, is almost completely consumed during combustion, as it leaves behind an ash (calx) of very small mass; hence it was regarded as nearly pure phlogiston. Thus, phlogiston theory is a mirror image of our current oxidation theory. For the case of metals, oxidation theory requires that you add oxygen to the metal to make the oxide, whereas phlogiston theory requires that you add phlogiston to the calx (oxide) to make the metal.

Most readers will be familiar with these fundamentals of the modern understanding of oxidation: Air is about one-fifth oxygen and four-fifths nitrogen. Oxygen is a highly reactive gas, whereas nitrogen gas is quite inert. Heating metals such as iron or copper or mercury in air allows them to *oxidize*—that is, form molecular bonds with oxygen. The oxides look completely different from the pure metal, just as rust looks different from iron. If enough metal is present in a heated sealed vessel to react with and totally exhaust the available oxygen, then only nitrogen (and about 1% of the total volume residual argon) will remain. Nitrogen will not support combustion or respiration by animals because both of these processes require oxygen. Chemists denote reactions that add oxygen to atoms or compounds as *oxidation* and those that reverse this process to remove oxygen atoms (or, in general, add electrons) as *reduction*.

Here is one reason why phlogiston theory was so doggedly accepted: It neatly but incorrectly explains why it is necessary to burn charcoal (or coke, its cousin) with some strongly bound oxides, including iron ore, in order to produce the pure metal. In phlogiston theory terms, adding extra phlogiston (charcoal) allows the calx (iron ore) to convert to the fully phlogistonated (metallic) form. In modern terms, the explanation is unfortunately more complex: Oxygen atoms are so tightly bound to the iron atoms in iron ore that the iron cannot be reduced (or torn away from oxygen) merely by heating to even white heat. The carbon atoms in charcoal, however, voraciously tear oxygen atoms from iron oxide when they burn to form the tightly bonded gaseous products CO and CO_2, leaving the metallic iron atoms behind [3].

But phlogiston theory had so many flaws that these few triumphs were insufficient. As eighteenth-century chemists developed airtight glass and metal retorts, they noticed that combustion of a substance like charcoal was incomplete if too small a quantity of air was available. However, air had not been required for combustion in the original form of the phlogiston theory. To patch up phlogiston theory, they hypothesized that sufficient air must be present to absorb all of the phlogiston to achieve complete combustion. It is characteristic of a dying theory that it spawns more and more suppositions.

In fact, an even more outrageous supposition was necessary to rescue this ass-backwards theory. Consider how phlogiston theory would apply to the case of burning iron filings. There is a painful complication here that would become evident to anyone taking the trouble to weigh the filings before and after burning. Whereas wood and charcoal clearly diminish in weight during combustion (presumably, as the phlogiston is forced out), iron filings increase in weight after they are burned. Oops! That must mean that phlogiston can have a negative weight. Of course, phlogiston had a positive weight when it is in charcoal, because charcoal's weight diminishes as it combusts and the phlogiston is driven out. So phlogiston can have either a positive or a negative weight. This theory was beginning to reek.

Although many chemists played important parts in explaining that air contains oxygen which has a key role in combustion, we will focus on two key rivals: Joseph Priestley, a middle-class Englishman, and Antoine Lavoisier, a French aristocrat. Not only were they the outstanding spokesmen for their dueling views, but also they knew each other personally and offended each other mightily—and their competition was enmeshed in the growing hostility between England and France.

Joseph Priestley (Figure 4.1) is hardly a household name in the twenty-first century, but he was enormously influential in science, politics, and religion in the late eighteenth century. In *The Invention of Air*, Steven Johnson highlights just what a pivotal figure Priestley was:

> *In their legendary thirteen year final correspondence, ... Thomas Jefferson and John Adams wrote 165 letters to each other. In that corpus, Benjamin Franklin is mentioned by name 5 times, while George Washington is mentioned three times .... Priestley, an Englishman who spent only the last decade of his life in the United States, is mentioned 52 times [4, p. xiv].*

Priestley, the son of a weaver, was born in 1733 in Yorkshire. He was raised by a Presbyterian mother until her early death and a devout Calvinist aunt thereafter. Although he rejected the stern notions of Calvinism as a youth, a deep piety led

Figure 4.1 Joseph Priestley ca. 1766.
Source: The Library of Congress.

him to study religion and become a minister. He shared with Newton a skepticism of the doctrines of the Anglican Church—particularly of the Trinity and the divinity of Jesus—and these beliefs marginalized him because the Test Acts made any dissenter from the Anglican Church ineligible for Oxford or Cambridge educations or public employment. Priestley was nevertheless able to acquire an excellent education at a dissenting academy.

A humiliating stutter made his first pastoral postings unsuccessful, so he turned to writing as a way to augment his family's meager income. Showing the innovative bent that would characterize his scientific work, Priestley published in 1761 his first really successful book, *The Rudiments of English Grammar*, a trailblazing and educationally subversive attempt to undercut establishment education's unproductive obsession with Greek and Latin and to systemize English. An interest in science led him to befriend Benjamin Franklin and other "electricians" and to publish in 1767 an extremely influential *History and Present State of Electricity* that incorporated many of his own experiments. By 1768 he had entered the political arena and made himself a target for the establishment with a plea for civil liberties in his *Essay on the First Principles of Civil Government*. His interest in "airs" was piqued when he was posted to a Presbyterian church in Leeds next to a brewery where he began experimenting with the CO_2 evolved during fermentation. He soon invented carbonated water.

Priestley's wide-ranging experiments in this period were vital in illuminating the complex nature of gases. Working with inexpensive crockery and glassware borrowed from his wife's kitchen (Figure 4.2), he discovered hydrogen chloride, ammonia, sulfur dioxide, and nitric oxide. The latter may be easily generated by

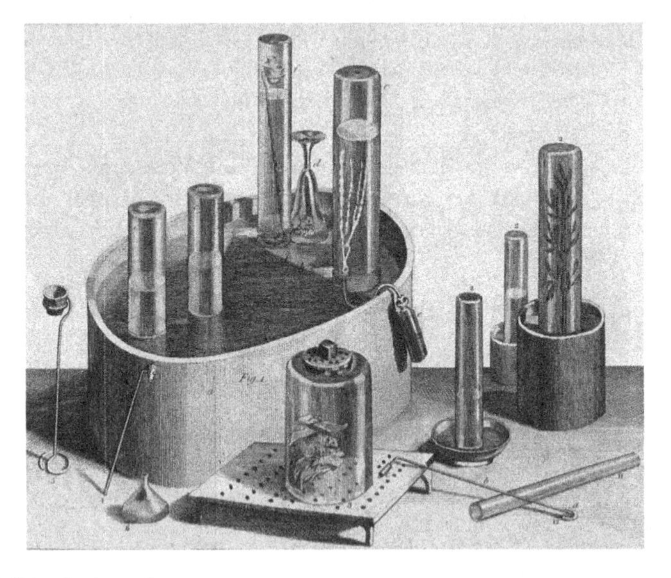

Figure 4.2 Priestley's equipment.
Source: The Library of Congress.

heating nitric acid with metals. Toying with the nitric oxide (NO), he found that if he mixed it with atmospheric (common) air he would produce reddish brown, water-soluble nitrogen dioxide (NO_2). By measuring the volumetric shrinkage of the common air, Priestley could calculate its fraction of respirable air (oxygen) [5, p. 111]. Priestley was delighted with this test and used it in many experiments to show that the most respirable air was found outdoors and that respirable air was depleted in an unventilated meeting room. These experiments first really verified that "air" is a mixture of at least two gases.

Here is another example of how fertile Priestley's conceptually simple experiments could be. Acting apparently on an impulse in 1771, he put a bare-rooted mint plant in a jar containing only water. He knew that animals require air for respiration, so he expected the mint to die, but day after day it not only survived, but also gave off colorless gas. Most curious! Plants do not require air to survive as animals do. Moreover, when he tested the gas produced by putting a mouse in it, the mouse survived [4, pp. 61–84]. The plant had somehow *produced* respirable air. Animals cannot survive in an atmosphere exhausted by a burning candle, he thought, but can plants? He set up an experiment in which the mint was exposed to a volume of air trapped with a burning candle, which he knew would exhaust the respirable air. The candle was soon extinguished, but the plant not only grew day after day but also amended the air in which it lived, making it once again respirable and able to sustain a mouse's life. Priestley grasped immediately the significance of this experiment: Plants can purify exhausted air and make it suitable for animal respiration, thus making him at once the founder of plant physiology and a grandfather of the twentieth-century environmental awakening.

Priestley repeated these experiments many times and performed a careful series of control experiments to confirm the incredible results he had observed: verifying that the candle's combustion products were not amended simply by the passage of time or by remaining in contact with the water. Although other plants, including balm and the evil-smelling groundsel also amended air, a spinach plant produced even more respirable air than mint.

This series of experiments shows a master scientist at work. Priestley, in effect, was having a slow but deliberate conversation with Nature. He submitted a question, and Nature answered. He deliberated and submitted a follow-up question. Nature deliberated and answered. Each answer is specified exactly by the experiment posed, but do not count on Nature to clarify the syntax. She will not give you any generalized or gratuitous information, yet she always replies precisely to your *specific* question.

Furthermore, consider also how simple and yet significant this series of experiments is. The materials were available in the ancient world, but the performance and interpretation had to wait millennia. It is not the tools that make experiments great—it is the tools in the hands of a master who can use each discovery as a lever to pry open the next locked door.

With these and other experiments, the brilliant Priestley danced all around the mystery of oxygen. He was like a foredoomed tennis player who makes it to match point again and again but is unable to claim his victory. In his mind, he was still removing phlogiston from air rather than producing oxygen. The assembly of this fragmentary knowledge into a broad and coherent picture had to await the attention of a more systematic scientific genius, the father of modern chemistry, Antoine Lavoisier.

Lavoisier was born in 1743 in Paris to a wealthy mother and a father who was a solicitor to the most powerful court in France. He was educated at the Collége Mazarin, the best school in Paris, where he received his baccalaureate degree in laws in 1763. However, during his last years of study he attended the lectures of Guillaume Rouelle and developed a passion for chemistry. Sensing the seriousness of his interest, geologist Jean-Étienne Guettard, a family friend, invited the promising young graduate to tour the Vosges and nearby Switzerland with him to map the geology and resources of this region, and the youthful Lavoisier gained a broad knowledge of minerals and chemical processes. After his return to Paris at the conclusion of this 4-month effort, Lavoisier published his first scientific work, describing how gypsum lost its water of crystallization on being heated for conversion into plaster of paris. Receiving a special award from Louis XV for his next paper on Paris street lighting, Lavoisier showed himself a prodigy by being elected the youngest member of the French Academy in 1768. This honor carried with it considerable responsibilities, and Lavoisier completed many investigations into the practical side of chemistry for the academy. Most notable were his improvements in the art of gunpowder manufacturing that made French powder supreme in the world and allowed the American colonists to outshoot the Redcoats.

But this young man in a hurry had financial aspirations too, as he demonstrated by purchasing a share in the Fermé Général, making himself a tariff collector eligible for a 10% share of the duties levied. Lavoisier completed his transition to a prosperous and respectable adulthood in 1771 when he married the 14-year-old Marie-Anne Paultze. Marie became a beautiful woman, a beloved wife, a treasured lab assistant, an indispensable translator of English scientific articles, the illustrator of Lavoisier's scientific papers, and the grand dame of the most sophisticated scientific salon in Paris (Figure 4.3). So, despite being born with a silver spoon in his mouth, Lavoisier pursued his career and affairs with diligence and grace. If you did not know the conclusion of the story, you might even be tempted to envy him.

Lavoisier's first significant work was to determine whether water could create earth. After all, the evidence would have been clear to anyone who ever boiled down a pot of water. When the water disappeared, the sediment of various types remaining in the bottom of the pot could have been engendered by the water—that is, water begat solid material. The first chapter of Genesis might even be read to mean that the dry land God commanded to appear from the original water was produced from the water. Furthermore, other traditions around the globe, including the writings of the ancient Greek philosopher Thales, also taught that water begat earth.

A well-known and highly influential experiment furthering this view was performed by Jean Baptista van Helmont and published in 1648, 4 years after his death [6]. He had planted a 5-lb willow tree in a pot containing 200 lbs of dried soil and provided only distilled water or rainwater to nourish the growing tree, minimizing

Figure 4.3 Antoine and Marie at work. *Source:* 1788 portrait by David, now at the Metropolitan Museum of Art.

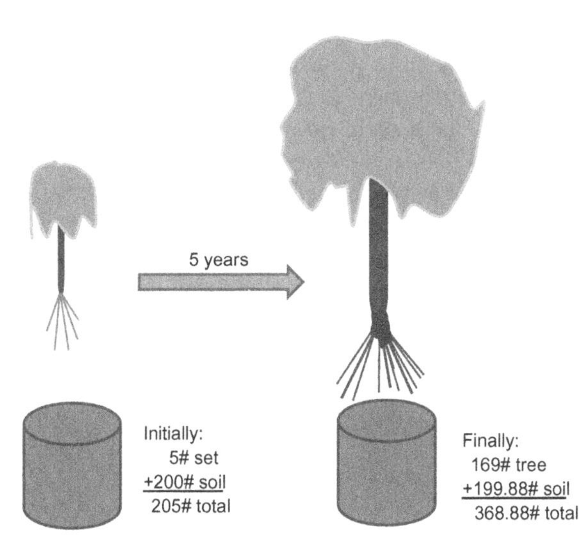

Figure 4.4 Van Helmont's 5-year willow tree growth experiment.

dust deposition in the pot by covering it with a perforated lid (Figure 4.4). After 5 years, the bare-rooted willow alone had grown to 169 lbs, whereas the dried soil was found to have lost only 1/8 lb. The weight of the leaves that fell in four autumns was not included in the experiment. Van Helmont concluded that the added 164 lbs of willow "arose out of water only."

A modern plant physiologist would counter that much of the weight increase of the growing willow was indeed water, with about one-fifth of the wood's weight, the dry weight, being hydrocarbons extracted from carbon dioxide metabolized from the atmosphere and hydrogen and oxygen extracted from the water. One might expect that 1% of the weight of the willow might have come from minerals extracted from the soil, which should have depleted the soil by 1.6 lbs rather than 1/8 lb. The discrepancy might be easily attributed to the difficulty in drying the soil twice to the same level of residual water or experimental balance error.

Lavoisier, who was not convinced of the relevance of van Helmont's experiment to chemistry, devised his own experiment to show that water did not create earth. In 1768 he placed 1.468 kg of eight times distilled water in an airtight pelican (Figure 4.5) and heated it to near boiling for more than 3 months [7]. When heated from below, the upper arms of this flask allow the boiling water to condense and drain again to the bottom. After 101 days, the airtight pelican was cooled and opened. Do you have the feeling that only a scientist who could afford many servants could accomplish 101 days of careful temperature control? Although the weight of water in the pelican remained unchanged, 1.00 g of thin, earthy flakes were found in the bottom of the pelican, whereas the pelican's dry weight had diminished by 0.82 g, presumably as the aforementioned flakes spilled from the glass surface. The overall discrepancy implied a weight gain of about 1 part in 10,000, which Lavoisier attributed to experimental error.

With its emphasis on nearly boiling water, this experiment seems to have been designed to address the question of how residual material at the bottom of a pot is formed as water boils. Had the experiment been performed at room temperature with traces of photosynthesizing single-celled plants present, Lavoisier might have seen a mass of algae formed. As it was, his experiment did just about all that the technology of the time could do to show that sustained heating of water could not produce solid matter. Lavoisier would have had to maintain the pressure in the pelican at below a few atmospheres to avoid breaking the seal, so he would have had to exhibit some care not to heat the sealed sample much above the boiling point at 1 atmosphere. The experiment confirms Lavoisier's commitment to using the

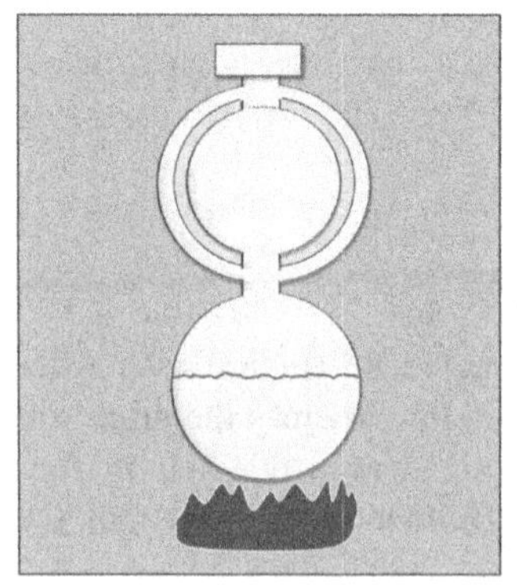

Figure 4.5 A pelican flask showing water level and condensing side arms.

conservation of matter to unravel chemical processes while it demonstrated just how accurately this principle could be applied.

This neat little experiment shows Lavoisier's genius beginning to flower. Ignoring all the medieval notions about phlogiston and the caloric fluid, this experiment demonstrated that mass conservation could be the tool to unravel chemistry's deepest mysteries. Consider how he simplified the van Helmont experiment:

- Recognizing that a living organism introduced a whole new set of imponderables, he avoided using a plant.
- He took the experiment into the laboratory, where it belongs, and avoided the problems of wind-borne dust or insects.
- He reduced the mass of the material to be weighed from 200 to 3 lbs. This improves the accuracy and minimizes the problem of drying the soil, which may be inadvertently burned or volatilized.

You can imagine how such a quantitative success would have buoyed Lavoisier's confidence. Delayed because of his time-consuming obligations to the academy, he eventually set up a series of experiments to attack the vulnerabilities of phlogiston theory. In 1772, he investigated the burning of phosphorus, finding in agreement with Cigna and Guyton that phosphorus increased in weight after burning [2, pp. 102–103]. Lavoisier was beginning to believe that something in the atmosphere might be combining with the phosphorus during combustion. Further experiments showed weight gain as sulfur and lead burned also. A nicely conceived attempt to explain a 100-year-old mystery of chemistry—the disappearance of diamonds when heated to high temperature—failed because the massive burning lens cracked the glass vessel in which the diamond's combustion products were to be collected [8].

Although by 1774 the structure of phlogiston theory was proving decrepit and rickety, the phlogistonists had erected a sturdy wall of protection. They had long ago proved themselves resistant to criticisms that should have dealt their theory a fatal blow. After all, Gabriel Venal had written in the third volume of Diderot's *Encyclopédie* that instruments and "artificial measurements" had no place in their work. Whereas the physicist investigated the gross properties of bodies by calculation and measurement, the chemist sought the "elusive inner properties of matter, accessible only by methods which were indirect and intuitive" [9]. This reads like a determined attempt to remove all of chemistry from the realm of science.

Meanwhile, Priestley also had been prospering and refining his techniques. His success as both a teacher and a scientist attracted the attention of the Earl of Shelburne, who along with Priestley sympathized with the cause of the American colonists. Shelburne found Priestley a perfect fit as a tutor for his children and household librarian and hired him for both his political and his scientific acumen. Light duties, a generous salary, and lodgings on the Shelburne estates allowed Priestley to publish five books on the preparation and properties of "airs" during this period of employment. In 1774, he used Lord Shelburne's generous salary to purchase a 1-foot-diameter "burning glass" to focus the sun's rays on an expensive sample of red calx of mercury (mercuric oxide). Mercuric oxide was difficult to prepare but in great demand as a treatment for syphilis. Since a whole series of chemists from Boyle through Scheele had reported the release of gas from heated red calx, Priestley set the bell jar containing his precious sample over a trough of mercury so as to collect any gaseous products. As Priestley adjusted the burning glass's wooden frame to focus solar heat on the calx, it released volumes of colorless gas. When gaseous evolution ceased, Priestley thrust a candle into the trapped gas in an effort to determine whether it was "memphetic air" (nitrogen), which he expected would extinguish the candle. In an unexpectedly profound moment for science, the candle did not go out but burned more brightly than it did in air and with an enlarged flame [5, p. 126]. Intrigued, he experimented further, finding that a chip of red-hot wood burst into vigorous flame and a hot wire glowed white as the sun when thrust into the unknown gas. Priestley had discovered what his rival Lavoisier would christen "oxygen," although Priestley at the time incorrectly suspected it was nitrous oxide, now known as "laughing gas."

Priestley's experiments were interrupted at this time as he prepared for a tour of Europe with Lord Shelburne. During a visit to Paris, the two were guests of honor at a lavish dinner attended also by Antoine and Marie Lavoisier [5, pp. 158–162]. Priestley, direct and trusting as always, talked extensively and explicitly about how to produce from the calx of mercury the new gas that supported combustion better than atmospheric air.

After his visit to Paris, Priestley continued mulling over the red calx experiments; it began to gnaw on him that the gas he had discovered appeared to be more perfect for respiration than atmospheric air. How did this fit his own certainty that God had tailored an optimal world for occupation by his children?

Returning to England, Priestley continued his experiments with this mystery gas, following up his hypothesis that it was "diminished nitrous gas" or nitrous oxide. Many unsuccessful attempts to dissolve the gas in water proved that it was not nitrous oxide. The baffled Priestley put the experiment aside for 3 months, finally returning to test the respirability of his new gas with mice. He found that a mouse placed in his new gas could live twice as long as in common air. Ultimately Priestley even tried breathing it himself, finding no ill effects and reporting that afterward "my breast felt particularly light and easy for some time afterwards" [4, p. 89]. Priestley had done enough experiments; now was the time to stake his claim. In 1775, he submitted a report to the secretary of the Royal Society that he had made "dephlogistonated air," thus missing the opportunity to lay claim to one of the great discoveries of the eighteenth century.

By early 1776, Lavoisier had mulled over the significance of Priestley's red calx of mercury experiment enough so that he was ready to improve on it and perform his most brilliant experiment, one that would drive a stake through the heart of the phlogiston theory. His apparatus was a long-necked retort that he called a *matrass*, which was bent as in Figure 4.6 so as to extend under a partially evacuated bell jar placed in a trough of mercury [10, part 1, ch. 3]. He placed 4 ounces of pure mercury in the matrass and lit the furnace to heat the mercury within the matrass to nearly boiling. At first, no change was visible, but by the second day small red particles of calx of mercury (mercuric oxide) began to form on the

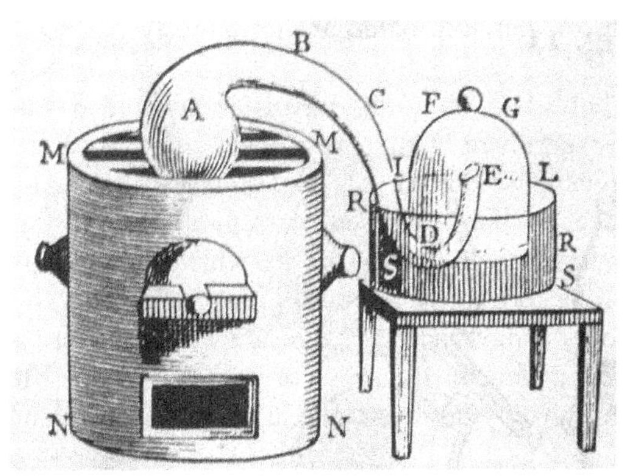

Figure 4.6 Lavoisier's apparatus for oxidation of mercury.
Source: Drawing by Marie Lavoisier, *Elements of Chemistry*, Plate 3.

mercury's surface. They increased in size and became more numerous until it became clear by the 12th day that new material ceased forming. After the fire was extinguished, the collected calx weighed 45 grains (2.92 g). At the same time, the original 50 cubic inches of air in the matrass had diminished to "42 or 43 cubic inches" as the oxygen was depleted; moreover, the residual gas did not support combustion or respiration.

But Lavoisier was not finished with the calx. He wanted to reverse the reaction to regain the original reactants. He therefore placed the scraped calx in a smaller matrass similarly arranged with its neck in a mercury trough and its mouth under a bell jar and then heated the calx red hot. In a few minutes, the calx decomposed, leaving behind pure mercury and "7 or 8 cubic inches" of gas that was more capable of supporting combustion or respiration than air. The ratio or fraction of air that is respirable would be about 16% if calculated from this experiment, but Lavoisier recognized that not all of the respirable air formed calx in the original experiment; some unreacted fraction remained mixed with the memphitic air. In modern terms, the oxidation of mercury did not go to completion. Lavoisier's description in his *Elements of Chemistry* refers to other experiments that show "at least in our climate that the atmospheric air is composed of respirable and memphitic air in the proportion of 27 and 73" (twenty-first-century measurements: oxygen 21%, nitrogen 78%, and argon 1%).

The alert reader may be wondering why heating mercury near its 357°C boiling point in air can form mercuric oxide, whereas heating to a red heat—more than 800°C—rapidly decomposes the same oxide. The low-temperature oxidation reaction proceeds because the 3 lbs abundance of mercury provides many opportunities for oxygen atoms to form mercuric oxide. Moreover, as Lavoisier noted, the reaction never completes in the sense of using up all of the oxygen in the matrass anyway. Conversely, during the decomposition of mercuric oxide, little mercury but much mercuric oxide occupies the matrass, pushing the reaction toward mercury and gaseous oxygen. Another factor encouraging the mercuric oxide to decompose, particularly at high temperatures, is that oxygen atoms bound as solid oxides release much energy when they form gaseous O_2. This reversibility is not true for every oxide, but the thermodynamics of mercuric oxide are fortuitously balanced so that it is possible for mercury.

In his *Elements of Chemistry*, published in 1789, Lavoisier concluded the description of this groundbreaking experiment with a somewhat more accurate approximation than that originally given by the red calx experiment: "the air of the ancients is not a simple substance ... but composed of respirable air to the extent of one quarter and that the remainder is a noxious gas ... which cannot support combustion." He still maintains one foot in the mystical past of chemistry, as he struggles on for a few hundred words about when the "caloric and light must be disengaged" as the mercuric calx forms. Fire, in the sense of heat and light, still seems like one of the elements, but fortunately not one that exhibits enough weight to invalidate his historic experiment.

Again and again in science, a confused corpus of observations and tenuous theories is suddenly illuminated by one blindingly simple experiment or system. For instance, the Galapagos Islands provided Darwin a simplified world in which he could more readily disentangle the effects of evolution. Lavoisier's experiment with the red calx of mercury is one such beautifully simple system. Priestley had discovered that mercuric oxide could be reduced to make oxygen and Lavoisier made it oxidize again while keeping track of the fraction of atmospheric air it consumed. Lavoisier had finally shown quantitatively that something in the air combined with mercury when it oxidized and could be driven out again on high-temperature reduction.

A few pages after describing this experiment in *Elements of Chemistry* (1789), Lavoisier [10, part 1, ch. 4] baptized respirable air with our current name, *oxygen*, from the Greek for "acid former." The new baby was finally christened, and the first steps toward a scientific revolution were documented.

It is churlish to quibble with genius, but this nomenclature could be classed with a long tradition of imperfect scientific name selections, as it is really hydrogen that is more characteristic of acid formation. However, the name *oxygen* has not generated as much confusion as Ben Franklin's assignment of "vitreous electricity" as "positive," which a century later turned out to give the primary constituent of electricity, the electron, a negative charge. Franklin's nomenclature requires, for instance, that a volume of a semiconductor having a deficit of electrons has a positive charge.

As for the phlogiston theory, Lavoisier published its most pungent rebuttal in 1785 [2, p. 111]:

> *All these reflections confirm what I set out to prove [in 1783] and what I am going to repeat again. Chemists have made phlogiston a vague principle which is not strictly defined and which consequently fits all the explanations demanded of it. Sometimes it has weight, sometimes it has not: sometimes it is free fire, sometimes it is fire combined with an earth; sometimes it passes through the pores of vessels, sometimes they are impenetrable to it. It explains at once ... transparency and opacity, color and the absence of colors. It is a veritable Proteus that changes its form at any instant.*

Did the lucid exposition of *Elements of Chemistry* rapidly convert the phlogiston theorists? Only very slowly; in fact, this may be the best example of Kuhn's pessimistic view that a paradigm shift may not be complete until the old guard dies off. Priestley, his strongest opponent, felt that Lavoisier had not given him sufficient credit for his initial work with the red calx of mercury. If he were to accept Lavoisier's oxygen interpretation, then Lavoisier would owe him little priority for his experiments. If, on the other hand, the phlogiston theory were retained, then Priestley rightly deserved credit for being the first to produce and recognize "dephlogistonated air." Still indulging this personal agenda, Priestley went to his grave holding fast to phlogiston. In fairness, Lavoisier's recasting of chemistry, though a brilliant and seminal work, had left a number of major problems unresolved. How about this "caloric," or weightless energy that flowed out of reacting bodies to make fire? That sounds as mystical as phlogiston. The solution to this problem had to await the work of James Prescott Joule establishing the conservation of energy in 1835. So if one were stubborn enough to resist Lavoisier's chemical innovations, there was no shortage of valid criticisms.

Lavoisier was a truly revolutionary innovator. He was widely criticized, particularly in his early career, for failing to reference his predecessors, but in his own mind chemistry began with him. He was too skeptical to accept the frequently careless conclusions of others. He did not just reference the experimental contributions of earlier scientists—he repeated them. His perspicacity in assigning oxygen as an element cleared the way for him to clarify which other substances were elements. Lavoisier utilized the operational definition that elements were substances that could not be further refined, decomposed, or reduced. Clearly the 17 known metals in columns 3 and 4 (Table 4.1): iron, mercury, gold, silver, and so on, now unencumbered with phlogiston, could be designated as elements. The gases oxygen, azote (nitrogen), and hydrogen were also elements. As you can see, he was, at one stroke, well on the way to assembling the periodic table of elements. The very strongly bound oxides such as alumina and silica that he could not reduce had to tentatively be considered elements, but publishing such a list did a great service to chemistry because it gave future researchers a ready target to aim at.

Table 4.1 Lavoisier's Table of Elements [10, part 2, section 1]

Light	Sulfur	Antimony	Mercury	Lime
Caloric	Phosphorus	Arsenic	Molybdena	Magnesia
Oxygen	Charcoal	Cobalt	Nickel	Barytes
Azote	Muriatic radical	Copper	Platena	Argilla (alumina)
Hydrogen	Fluoric radical	Gold	Silver	Silex (silica)
	Boracic radical	Iron	Tin	
		Lead	Tungsten	
		Manganese	Zinc	

In the next few years, Lavoisier vigorously politicked for the acceptance of his new theory. The phlogostinists had editorial control of the major French scientific journal, *Journal de Physique*, so he founded *Annales de Chimie* as an outlet for anti-phlogiston articles. He mailed gift copies of *Elements of Chemistry* to influential chemists in France and abroad and converted more and more scientists away from phlogiston and toward oxygen. But the sands of time were emptying Lavoisier's hourglass.

Reading how the Reign of Terror of 1793, the response to the stabbing of Jean-Paul Marat in his bathtub by Charlotte Corday, ensnared Lavoisier is dispiriting. The more realistic and decisive members of the aristocracy were fleeing: the Marquis de Lafayette had left in the previous year, Pierre du Pont had fled, and the philosopher the Marquis de Condorcet was in hiding. The mathematician and physicist the Marquis de Laplace disappeared. But whether from egotism, naiveté, or inertia, Lavoisier, the most brilliant chemist the world had ever produced, made a series of decisions that would prove fatal.

In November 1793, the second month of the terror, the revolutionary government determined to arrest all members of the Fermé Général. An incorrect address on his warrant allowed Lavoisier some breathing room. At first he wandered the streets of Paris, then he hid in the shuttered Academy for 4 days while writing notes to the authorities, citing the valuable work he had done for the government. Lulled by the government's promise that the Fermiers would be released after they settled their accounts with the government, Lavoisier turned himself in. After 6 months of imprisonment and a hasty trial during which his request for a delay of execution to finish his work on human respiration was summarily rebuffed with "The Republic has no need of scientists," Lavoisier was tumbreled to the guillotine.

In a less-direct way, Priestley also fell victim to the French Revolution, as his known sympathies for the French and American Revolutions and his Unitarian religion had made him the Jane Fonda of England (Figure 4.7). Lord Shelburne was a fair and generous man but still had political ambitions, so he had retired the hot potato Priestley with a small annuity in 1780. Priestley thus continued both his research and his preaching at the New Meeting Church in Birmingham. In 1791, with English fears over the godless French Revolution reaching a boiling point, a Birmingham mob burned his home and laboratory and drove Priestley to flee to London and ultimately to settle in Pennsylvania in the United States.

It turned out that Lavoisier did not have to worry about the long-term effects of heavy metal poisoning, but this story of scientists working with mercury illustrates once again how frequently those who use novel and exotic materials take extraordinary risks. What would the US Occupational Safety and Health Administration say about spending many days in an ill-ventilated room stooped over a trough of mercury? Even worse, high-temperature decomposition of mercuric oxide could release a toxic dose of mercury rather quickly if the flask's seal failed. In fact, several later chemists did succumb to mercury poisoning. Early in his career, Lavoisier had barely escaped a fatal explosion in his gunpowder works when he experimented

Figure 4.7 An anti-Priestley caricature from 1791 shows him trampling on the scriptures.

with replacing potassium nitrate with (what we now know to be) unstable potassium chlorate.

This tale underscores how much the very human enterprise of scientific research can be dependent on personality. Priestley was ever a dilettante, although a hugely talented and successful one, publishing original contributions to language, education, theology, electricity, and chemistry. Although Priestley was well positioned to be the discoverer that air contained a new element responsible for respiration and oxidation, he did not make this vital conceptual leap. After Lavoisier did perform this synthesis, perhaps Priestley felt that his contributions would be minimized as just more experiments verifying Lavoisier's conclusions if he acquiesced to the oxygen theory. For whatever reasons, Priestley adhered to phlogiston theory until his death in 1804.

The insider Lavoisier felt the urgency of building a new structure; he had a sincere compulsion to tear the damaged bricks out of the edifice of chemistry and rebuild it solidly for future ages. He wrote *Elements of Chemistry* that not only laid out his theory of oxygen but also contained Marie's extensive illustrations of the laboratory equipment he thought necessary for chemists. He founded a journal and politicked scientists around Europe and America to spread the new theory of oxygen. He realized that discovery alone would not be enough to put the new theory on solid ground; he had to proselytize also. Ultimately, his careful quantitative work and thoughtfully designed apparatus transformed chemistry into a quantitative science.

References

[1] Plato, in: E. Hamilton, H. Cairns (Eds.), Plato: Collected Dialogues, Princeton University Press, Princeton, NJ, 1987, p. 1526.
[2] W.H. Brock, The Norton History of Chemistry, Norton, New York, NY, 1993.
[3] Rayner-Canham, Overton, Descriptive Inorganic Chemistry, fourth ed., W.H. Freeman and Company, New York, NY, 2006, pp. 534–535.
[4] S. Johnson, The Invention of Air: A Story of Science, Faith, Revolution, and the Birth of America, Riverhead Books (Penguin), New York, NY, 2008.
[5] J. Jackson, A World on Fire, A Heretic, an Aristocrat, and the Race to Discover Oxygen, Viking, New York, NY, 2005.
[6] W. Pagel, Joan Baptista van Helmont: Reformer of Science and Medicine, Cambridge University Press, Cambridge, 2002.
[7] J.P. Poirier, Lavoisier: Chemist, Biologist, Economist, University of Pennsylvania Press, Philadelphia, PA, 1993, p. 32.
[8] G. Johnson, The Ten Most Beautiful Experiments, Knopf, New York, NY, 2008, pp. 45–48.
[9] C.E. Perrin, Research traditions, Lavoisier, and the chemical revolution, Osiris 4 (1988) 53–81 (2nd series).
[10] A. Lavoisier, Elements of chemistry, in: R.M. Hutchins (Ed.), Great Books of the Western World, vol. 45, Encyclopedia Britannica, Chicago, IL, 1952.

Bibliography

W.H. Brock, The Norton History of Chemistry, Norton, New York, 1993.
Detailed and technically excellent.
J. Jackson, A World on Fire: a Heretic, an Aristocrat, and the Race to Discover Oxygen, Viking, New York, 2005.
Vividly told!
G. Johnson, The 10 Most Beautiful Experiments, Knopf, New York, 2008.
The chapter on Lavoisier is guaranteed to keep your attention.
S. Johnson, The Invention of Air, Riverhead Books (Penguin), New York, 2008.
A. Lavoisier, *Elements of Chemistry*, in Great Books of the Western World, ed. R. M. Hutchins, Encyclopedia Britannica, Chicago, 1952, Vol. 45.
A real classic.
J.P. Poirier, Lavoisier: Chemist, Biologist, Economist, University of Pennsylvania Press, Philadelphia, 1993.

5 Michael Faraday Discovers Electromagnetic Induction but Fails to Unify Electromagnetism and Gravitation: It Is Usually Productive to Simplify and Consolidate Your Hypotheses

Michael Faraday will be our only Horatio Alger story, because, unlike the overwhelming majority of influential scientists, he rose from a family that was distinctly lower class. Michael was born in 1791 to a blacksmith father in such poor health that he could barely feed his family. The Faraday family, including two older children, moved to a desperately squalid area near Manchester Square in London in the mid-1790s in hopes of securing more work. What distinguished the Faradays from other poor families of their neighborhood was that they were adherents of the Sandemanian Church, a disciplined fundamentalist Christian sect that nurtured community and gave Michael his conviction that understanding the "book of nature" was as worthwhile as reading the Bible. They were a clean-living, close-knit, and endearing little group.

After a rudimentary education, Michael was apprenticed to bookseller and binder George Ribeau at age 13. The demanding craft of binding volumes in leather developed his dexterity and gave him the opportunity to study the texts that came to him for binding. He was particularly inspired by the developing glamour science of chemistry, so he bought a four-volume introduction that he disassembled and rebound with blank pages interspersed with the text for notes. At age 18, Michael began the practice of attending the one-shilling lectures of the City Philosophical Society and carefully taking notes to be bound and preserved. Ribeau, a kindly master, took pride in the notes Michael bound, showing a volume of them to a customer, Mr. Dance. Dance's father, William, was so impressed with Michael's knowledge and industry that he gave him tickets to the far more prestigious lecture series given by Sir Humphry Davy at the Royal Institution [1]. Davy was the scientific lion of early nineteenth-century England; this accomplished chemist and brilliant lecturer had become the heartthrob of London society (Figure 5.1).

How the Great Scientists Reasoned. DOI: http://dx.doi.org/10.1016/B978-0-12-398498-2.00005-9

Figure 5.1 Sir Humphry Davy. Engraving after a portrait by Sir Thomas Lawrence, 1830.
Source: The Royal Institution.

By 1812, Michael's 21st birthday and the end of his apprenticeship were approaching. Having given his heart to science, he simply could not endure a career as a bookbinder. He *knew* he had much to give as a scientist, but he felt the clutches of the nineteenth-century English class system tightening around him. However, like all Horatio Alger heroes, Michael now got his lucky break: A laboratory explosion temporarily blinded Humphry Davy, and William Dance arranged for Faraday to assist him as a secretary. Faraday performed well in this temporary post but was soon relegated to the book bindery again as Davy's eyes healed. In desperation, Faraday sent Davy a volume of the notes he had taken at Davy's own lectures and bound. Davy was impressed, and 3 months later was able to secure a post for Michael as laboratory assistant at the Royal Institution.

Faraday loved his first assignments, particularly the excitement of working with Davy on explosive nitrogen chloride. Even the routine task of extracting sugar from sugar beets was a delight to him. Davy must have seen right away what a treasure he had in Faraday, for when he planned a European grand tour with his new aristocratic bride, Jane Apreece, to accept the Napoleon Prize in Paris, he brought Michael along as a scientific assistant. At the last moment, Davy's valet canceled, and Faraday agreed to serve as a temporary valet also until a replacement could be engaged in France. Unfortunately for Faraday, a suitable replacement was never found.

Europe opened Faraday's eyes to the larger scientific and cultural world. After arriving in Paris, Faraday assisted Davy in identifying a substance that the savants of France had been unable to analyze [2]. That a French delegation, including the young André-Marie Ampére, would come to an Englishman for help in identifying an unknown product refined from gunpowder was remarkable because the Napoleonic wars were approaching a crisis and gunpowder was the enriched uranium of the early nineteenth century. After 10 days of smelly and explosive experimentation in his hotel room using his traveling chemistry kit, Davy verified that the unknown violet crystals constituted a new halogen. He christened it *Iodine*.

The Davy party ultimately left Paris to make their way through the Maritime Alps to Italy. In Rome, Faraday had free time to see the sights, although he assisted Davy in identifying the constituents of ancient pigments from Herculaneum and clarifying the chemistry of chlorine dioxide.

The arrogant Lady Jane bitterly resented Faraday's acceptance in the scientific world despite his status as her inferior. Always condescending, she never missed an opportunity to mortify him. Davy himself was kind to Faraday, but he continued for decades to presume on the master–servant relationship they had established to ask Faraday to run errands for him. Faraday bore these slights and insults with good grace, if not with total Christian forgiveness.

On their return to England after 18 months abroad, Faraday was promoted to superintendent of the apparatus and assistant in the laboratory and mineralogical collections at the Royal Institution; his career now safely launched. Over the next few years he grew in chemical skills, assisted Prof. William Brande's lectures at the Royal Institution, and gave his own lectures on inorganic chemistry at the City Philosophical Society. A portrait of a more mature Faraday is shown in Figure 5.2.

The Royal Institution had been established by Joseph Banks, president of the Royal Society, in 1799 as an "Institution for diffusing the knowledge, and facilitating the general introduction of useful mechanical inventions and improvements; and for teaching, by courses of philosophical lectures and experiments, the application of science to the common purposes of life" [3]. The Royal Institution was a secure base for Faraday throughout his life, but at the same time it bound him to the less-creative applied research that supported the Institution yet diminished his creative output. This is a very familiar problem for those of us employed in industrial research laboratories in the current era.

Figure 5.2 Michael Faraday, 1842. Portrait by Thomas Phillips.
Source: The National Portrait Gallery.

Faraday's first major scientific triumph stemmed from a chance observation in Copenhagen. While giving a lecture on electricity in 1820, the Danish physicist Hans Christian Oersted noticed that a current-carrying wire deflected a magnetized needle. His initial and entirely rational expectation was that the magnetic force that perturbed the compass needle emanated radially from the current-carrying wire. Only 3 months later did his follow-up experiments show that the magnetic force was oriented around circles normal to the wire (Figure 5.3). Such a force was completely novel, as both known forces, the gravitational and the electrostatic, acted radially toward a mass or toward or away from a charge. Oersted's publication created a sensation among scientists [4].

But why did it take the scientific world so long to discover the coupling between electricity and magnetism? The magnetic needle compass had been widely used in Europe since the fourteenth century, and Volta invented the battery in 1800 [1, p. 25]. We have to suspect that the novel orientation of magnetic force contributed to its tardy discovery.

One of the joys of the scientific life is participating in the heady excitement when revolutionary discoveries buzz through the research community. In London, Davy, who had recently been elected president of the Royal Society, barged into Faraday's lab with an English translation of Oersted's article. With Faraday as technician, the two reproduced Oersted's experiment line by line from the text, marveling at the unexpected transverse orientation of the magnetic force. Apparently, Davy did not find in Faraday the inspiring collaborator he wished, because in the next few months he began a series of discussions with William Wollaston, his fishing buddy and the inventor of a process for purifying platinum. It is a regrettable feature of the human mind that, faced with novelty, we frequently try to put the new wine into old bottles, and the imaginative Wollaston devised an unlikely way of reconciling transverse magnetism with radial electrostatic force. He hypothesized that the electric fluid sloshed in helical fashion through the wire, thus superimposing a helical symmetry on the magnetic force. Wollaston had expected that a current-carrying wire should spin about its own axis under the influence of an applied magnetic field, an effect that has since been demonstrated to be extremely small. The two friends tried some experiments that did not support this flight of fancy and let the matter lie. Although Faraday did not participate in the experiments, he was aware of this work.

Faraday, meanwhile, was not in a fit state of mind to capitalize on the potential of Oersted's discovery. During this period he was overburdened with developing

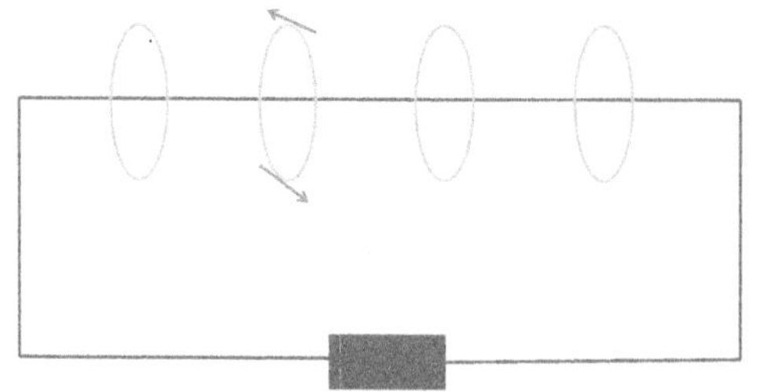

Figure 5.3 Oersted's experiment. Current flow from the battery through the wire establishes a magnetic field (light circles) that deflects the compass needle (arrow) along the magnetic field line in a plane perpendicular to the wire.

better steel alloys for cutlery, experimenting with lithography and photography, and improving the preservation of meat for the Admiralty. The hyper-responsible Faraday even penned a groveling letter of apology castigating himself for not meeting his contractual targets and vowing never to repeat this trespass. Another major distraction was that he was courting a fellow Sandemanian, Sarah Barnard, whom he married in the summer of 1821.

However, that summer the editor of *Annals of Philosophy* requested that Faraday write a review of the recent work on electromagnetism, and Faraday turned his attention once again to this enticing new topic. This rededication led to Faraday's first major discovery, the progenitor of the electric motor.

The circumstances of Faraday's discovery manifest again how invention is so frequently first implemented in toys. On September 3, 1821, Sarah's 14-year-old brother George came to spend the day in Faraday's laboratory. Faraday's mind, saturated with his reading on the new electromagnetic force, groped toward a way to help George visualize the transverse trajectory of the north pole of Oersted's compass needle. As inspiration began to strike, Faraday modified Oersted's experiment in an ingenious way: He turned the Oersted configuration inside out. Mechanically, it would be easier to spin a light wire around a heavy magnet rather than Oersted's configuration in which one's hand traces with a heavy compass needle the magnetic field oriented around a wire. Faraday could easily fix a bar magnet in place in a beaker with sealing wax, the all-purpose adhesive of nineteenth-century labs. All that remained was to pivot a wire above the magnet that could freely rotate while conducting current through the pivot and complete the circuit to the battery through a mercury bath (Figure 5.4). Michael and George were delighted to see the current-carrying wire begin to spin about its pivot and the magnet's long axis [2, pp. 79–81].

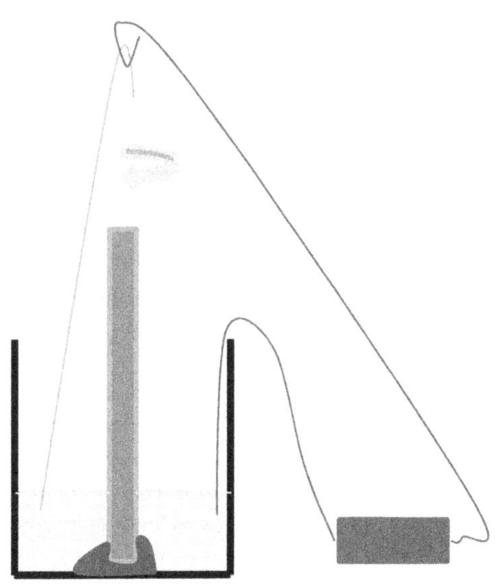

Figure 5.4 The first electric motor: Michael Faraday, 1821. The light-colored wire hanging from the pivot circles the bar magnet, which is held rigidly in the beaker by sealing wax. The battery supplies current through the pivot and the conducting mercury in the beaker.

George's boyish enthusiasm reinforced Faraday's glow of discovery; they circled the lab bench, mimicking the orbiting wire, and cheered. George later reminisced, "I shall never forget the enthusiasm expressed in his face and the sparkling in his eyes." Always scrupulous about documenting his experiments, Faraday then returned to his sober scientific persona to write a brief description of his hypothesis and apparatus [3, pp. 164–165]:

> *The effort of the wire is always to pass off at right angles from the pole, indeed to go in a circle around it. ... From the motion above a single magnet pole in the center ... should make the wire continually turn around. ... Very satisfactory, but make more sensible apparatus.*

This achievement deserved even more celebration, so Michael treated George to an afternoon at the circus. The afterglow of the morning's success stimulated Faraday to immediately publish this work to claim his provenance. In little more than a week he had submitted "On Some New Electro-Magnetical Motions, and on the Theory of Magnetism" for publication. He did not reference the helical current flow hypothesis of Wollaston or the failed experiments Davy and Wollaston had performed. As he prepared the paper, he described his successful motor with several colleagues, but not Davy, and his attempts to talk to Wollaston, who was not in London, were unsuccessful.

In his enthusiastic rush to emerge from the shadows to claim his first major scientific discovery, Faraday's prudence had failed him. It would in no way have compromised the momentous discovery he alone had just made to have generously acknowledged the unsuccessful attempts of Davy and Wollaston, but it would have saved him from the firestorm of criticism that now descended. Malicious rumors and press reports questioning whether Faraday had plagiarized key elements of his motor from Wollaston circulated rapidly through London's scientific circles. Reeling from attacks on his integrity, Faraday made a personal apology for this omission to Wollaston as soon as Wollaston returned to London. Wollaston did not appear to have been caught up in the fuss and accepted the apology graciously.

It is a reasonable guess that Davy was responsible for fanning the flames of this affair. Perhaps Davy was unable to accept that his former valet and technician was capable of such an intellectual achievement, as when a parent cannot update the capabilities of his maturing child, or perhaps he harbored a more culpable jealously at being surpassed by his own apprentice. At any rate, the relationship between Faraday and Davy was strained thereafter, and it is likely that Davy campaigned vigorously against Faraday and cast the only black ball when Faraday was elected to full membership at the Royal Institution in 1824.

Nevertheless, when Davy retired in 1825, Faraday's talents and achievements were so undeniable that he was appointed director of the laboratory under Brande as superintendent. Faraday was now able to institute regular Friday evening lectures on topics in chemistry and physics that soon attracted a rather glittering audience from London society (Figure 5.5) and cemented his reputation as a leading scientific light of England.

Figure 5.5 Faraday gives a Christmas lecture at the Royal Institution, ca. 1856.
Source: The Royal Institution.

The black memories of this electric motor affair must have nagged at Faraday's mind because he seems to have resolutely turned his attention away from electromagnetism and back to chemistry. In the next few years he developed a superior optical glass, discovered benzene, and was the first to successfully liquefy chlorine gas.

When he did return to electromagnetism briefly in 1825, we see him still trying to clarify the relationship between electricity and magnetism; in particular the problem of electromagnetic induction. For many centuries, people had induced magnetism in iron needles by stroking them with lodestone. Similarly, it had been known for decades that electrically charged objects could induce charges of opposite sign on conducting objects in their vicinity. Why had no one been able to induce an electrical current flow with a nearby magnet or current-carrying wire?

Ampére had demonstrated in 1820, almost immediately after Oersted's publication, that two parallel wires carrying current were either attracted to each other if their currents were parallel or repelled each other if their currents were in opposite directions. In 1825, Faraday looked for a related inductive effect, arranging two 1 meter long wires parallel to each other and separated by only the width of a piece of paper. He found that a current traveling through one wire did not generate measurable current in the parallel wire.

The diversity of Faraday's professional obligations may have slowed his progress in research, but they also gave him broadened perspective on the whole of physics. In giving some lectures with Wheatstone in 1828, Faraday devised experiments demonstrating acoustic sympathetic vibrations. These experimenters showed that when a metallic sheet was struck so that it rang, an identically sized metallic sheet several meters away could vibrate "sympathetically," ringing at the same frequency. Ideally, if both sheets were absolutely identical, or tuned to the same frequency, then sound waves from the struck object could transfer appreciable energy to the sympathetically resonating object.

These observations were thought provoking for Faraday because they hinted at a way around the most obdurate problem associated with Newton's 1687 theory of gravitation [5]. Newton's theory had been the most successful scientific idea ever; it had rapidly changed people's view of the universe from an unknowable mystery to a giant machine operating with understandable and simple rules. And the key rule was that every piece of matter in the universe attracted every other piece with a force proportional to the product of their masses and inversely proportional to the square of the distance between them. But how did masses far away from each other even sense each other's existence, let alone their direction or mass? Neither Newton nor any subsequent scientist had satisfactorily settled this issue, and it became known as the *action-at-a-distance* problem.

Faraday's demonstrations had shown that that energy could be transmitted by acoustic waves without any action-at-a-distance communication between the two resonators. Faraday turned this idea over in his mind for many years, wondering if it could point a way out of the puzzle of instantaneous communication between all masses. Could space be filled with waves that might transmit gravitational or even electrical or magnetic forces in the same way acoustic energy was transmitted? Faraday realized he would have to deepen his knowledge of electromagnetism before he could contribute a solid solution to this riddle, but he continued gathering and testing ideas that could bear on it.

Starting just a few years after returning from Europe, Faraday had been struggling with a memory problem that inexorably grew from a minor vexation to a complete nervous breakdown in 1839. Perhaps years of contact with toxic chemicals had been undermining Faraday's nervous system. Nevertheless, Faraday learned to cope with the memory loss by taking detailed and copious notes; Sarah assessed his mental state and dragged him off to Brighton for seaside vacations when she noticed telltale signs of his illness. Figure 5.6 is a photograph of the two together. By 1830, Faraday realized that one of the chief sources of stress in his life was his assignment to develop improved optical glass. This project had bogged down in a swamp of difficulties induced by impurities, inhomogeneities, and induced bubbles. This type of work demanded intense care and a lengthy course of patient experimentation yet offered only a low scientific payoff for Faraday. The plaudits would accrue to those exploiting the improved telescope mirrors and microscope objectives that would result from this arduous work. He decisively terminated this project by delivering a small sample of perfect glass to the Royal Society and flatly refusing to scale up the process for manufacture.

Figure 5.6 Michael and Sarah, ca. 1847. *Source:* The Royal Institution.

In the summer of 1831, Faraday was thus able to return his attention to electromagnetism once again. Very shortly thereafter he made a series of blockbuster discoveries.

Scientists had been puzzling for years over the nonreciprocal nature of the electric and magnetic forces. Oersted had demonstrated that a constant current of electricity excited a magnetic force that was strong enough to rotate a compass needle. However, many attempts to excite an electrical current in a wire by using a magnetic field had failed. A typical unsuccessful experiment was to place a length of wire connected to a galvanometer in a magnetic field or near a second wire carrying current and attempt to measure an "induced" current flow in the first wire, as in Faraday's unsuccessful experiment in 1826.

Faraday's first brilliant intuition of 1831 was to concentrate the effect of induction by working with a coil of wire rather than a straight length. He exploited this concept in his very first encouraging experiment as he wound two long coils of wire around a wooden block with the windings interspersed but insulated from each other [6]. By connecting one of the coils to his largest battery and the other to his galvanometer, he was able to measure a very miniscule pulse of current—but only when making and breaking contact with his battery. It was clear that he needed to increase the effect somehow.

His second breakthrough was to multiply the magnetic field by confining and intensifying it within a ring of easily magnetizable soft iron acting as a core to two distinct coils. Accordingly, Faraday had a ring of soft iron 2-cm thick and 15 cm in diameter forged. He wound two coils of 1.2-mm-thick copper wires around opposite sides of the ring (Figure 5.7), carefully insulated by twine laid between the adjacent strands of wire and calico cloth between layers. By making coil B with

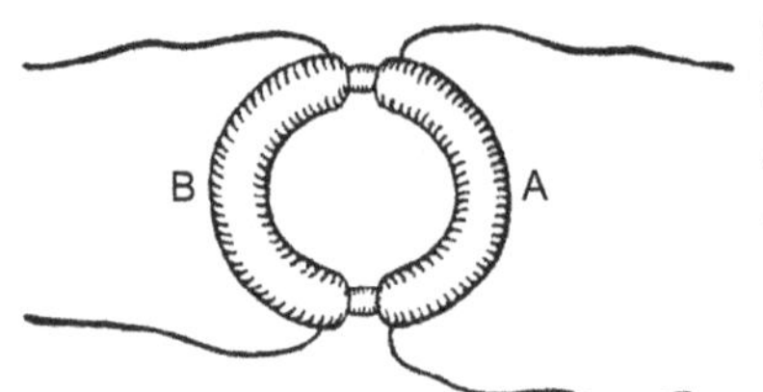

Figure 5.7 The induction ring, a prototype transformer, August 1831.
Source: Faraday's *Experimental Researches in Electricity*, Series I.

about 200 turns of wire, he would increase any induced voltage by a factor of 200. By inducing a magnetic field with coil A in soft iron rather than air, he would increase the magnetic flux within the ring by a factor of more than 1000. Altogether, Faraday had boosted the sensitivity beyond that of his 1826 experiment by a factor of 200,000. In performing the experiment, he connected coil B to his *galvanometer*, which at this stage of his research just consisted of a compass with an electric wire lying across it in a north–south orientation. He now attached coil A to his battery while he searched for an induced current measured by the galvanometer.

At the very second when he touched the battery to coil A, the galvanometer gave a lusty jump. Finally, real success! Yet not exactly the success he had hoped for. The galvanometer quickly dropped to zero after the initial rise and stayed there. Disappointed, he disconnected coil A from the battery. Coil B's galvanometer showed another lusty jump—but in the opposite direction. Faraday had just discovered electrical induction, the phenomenon that links electricity and magnetism, and fabricated the first electrical transformer. The rather disappointing observation that the induction was just a transient effect arose, of course, because transformers do not work with direct current.

Alas, Faraday was working alone as usual in his laboratory, so no eyewitness such as George has provided an account of his goal line celebration. It is reasonable to believe that Faraday initially regarded the discovery of induction as a limited success, as indicated by the dry account of what happened on connecting coil A to the battery in his daybook: "Immediately, a sensible effect on needle. It oscillated and settled at last in original position." After he repeated these experiments the next day, he left with Sarah by coach for a seaside vacation in Hastings. Despite their habitual but economical seats atop the coach and the pouring rain, Faraday was in such high good humor that he laughed aloud from time to time. As he later wrote to a friend, "I fancy my fellow passengers thought I had got something very droll in hand; they sometimes started at my sudden bursts . . ." [3, p. 247]. Later in the same letter he was unable to restrain his hopes: "I am busy just now again on Electro-Magnetism and I think I have got hold of a good thing but can't say; it may be a weed instead of a fish that after all my labour I may at last pull up."

Faraday was working in an era when the fundamentals of electricity were so poorly understood that he was not justified in assuming that magnetism developed by the current flowing in coil A was identical to the magnetism residing in a permanent magnet. His next step, therefore, was to prove that a similar inductive current would flow in a wire coil wound around an iron core magnetized by a permanent magnet rather than an electromagnet. On his return to London in late September, Faraday spent a frustrating day trying to reproduce the intriguing effects he observed in August with a permanent magnet rather than an electrical coil exciting the magnetic field. At the end of the day, showing a growing grasp of the essentials of his new discovery of induction, Faraday assembled the apparatus depicted in Figure 5.8 [6]. It amounted to permanent magnet forceps. In the closed position shown, a large magnetic flux traveled in a closed path through the two permanent bar magnets and the soft iron core of the coil. If the top leg was opened (broad arrow), the magnetic flux through the center of the coil decreased markedly. Mechanically opening and closing the magnetic circuit was functionally equivalent to turning the battery off and on in the induction coil depicted in Figure 5.7.

Just as in his experiment of August, Faraday found that opening the tweezers excited a movement of the galvanometer needle, and closing the tweezers gave a twitch of the opposite polarity. He wrote in his daybook, "Hence here distinct conversion of Magnetism into Electricity. Perhaps might heat a wire red hot here . . . ." Faraday implies here that he was extracting energy from the magnetic field rather than expressing our contemporary view that he was converting the mechanical energy required to swing the magnet open and closed into electricity.

Despite being uneducated in math, and, in reaction, somewhat disdainful of higher mathematics, Faraday had just succeeded in doping out the law of induction. It is best written in the language of differential calculus as

$$V(t) = n\frac{\mathrm{d}\varnothing(\mathrm{t})}{\mathrm{d}t}.$$

Here $V(t)$ is the voltage developed across the galvanometer at any time t, n is the number of turns in coil B connected to the galvanometer, and $\varnothing(t)$ is the time-dependent change in magnetic flux inside the magnetic core. The derivative $\mathrm{d}\varnothing(t)/\mathrm{d}t$ is a mathematical function that is positive and large when $\varnothing(t)$ is increasing rapidly, 0 when it is constant, and negative and large when $\varnothing(t)$ is decreasing rapidly. Figure 5.9 sketches the performance of these functions. Note that the derivative mimics the galvanometer signal Faraday observed, whereas the prefactor n shows how correct his intuition was that each turn of the detector coil B would add to the signal strength.

Faraday capped his work in late October by building a prototype electrical generator. To exploit as great a magnetic field as possible, Faraday traveled to

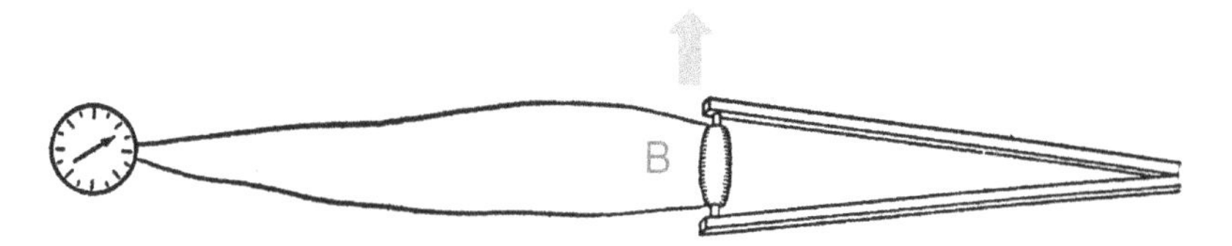

Figure 5.8 An induction generator, September 1831.
Source: Modified from a figure from Faraday's *Experimental Researches in Electricity.*

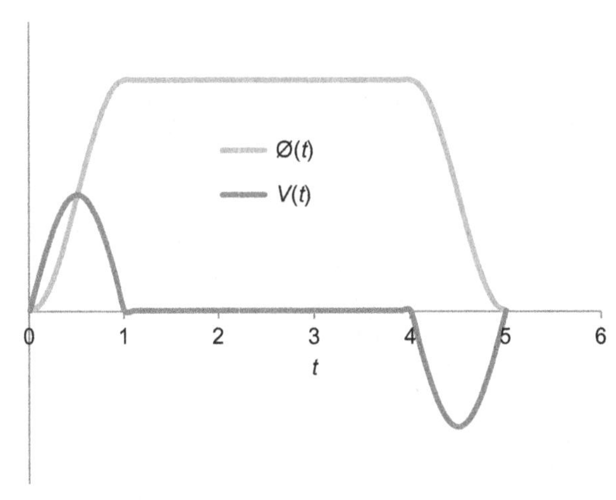

Figure 5.9 A comparison of the magnetic flux within the core of a Coil B from Figure 5.7 or 5.8, $Ø(t)$, and the induced voltage generated across the galvanometer coil, $V(t)$, as a function of time, t.

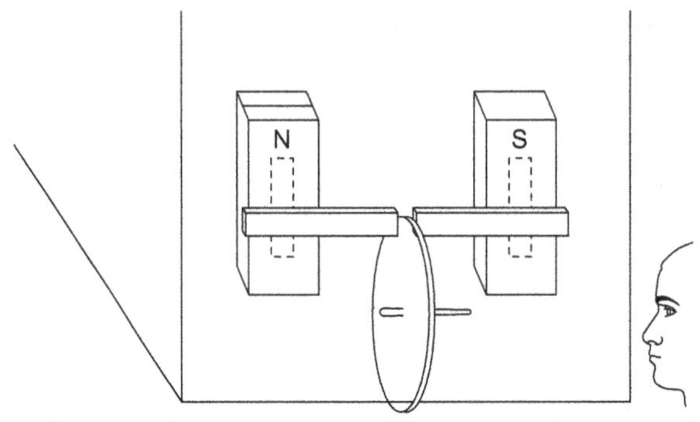

Figure 5.10 A prototype direct-current generator. *Source:* Faraday's own sketch of 1831 from *Experimental Researches in Electricity*.

Woolwich to use the great bar magnet capable of exerting 100 lbs of force maintained by the Royal Society there. He first verified that the larger magnetic field would increase the induced currents he had already produced by thrusting coils within the gap of the great magnet and observing larger galvanometer spikes than he had yet measured. With great anticipation, then, he installed the apparatus he had come to test: the first dynamo [6]. As Faraday's sketch (Figure 5.10) shows, he mounted a 30-cm diameter copper disk on a brass axle so that the disk's periphery could spin freely through the magnet's gap where the field was strongest. As Faraday spun the disk, he slid the two leads to a galvanometer over the surface of the smooth copper disk. To his great joy he discovered that, for the first time in 3 months of experimentation, he could finally detect a sustained voltage. The maximum voltage was obtained between the periphery near the magnet's gap and the axis of rotation.

Faraday had already developed a nascent theory of induction that could qualitatively explain these results. He believed that the induced voltage should be maximized in a direction both perpendicular to the magnetic field (directed across the magnet's gap) and the path of the conductor (around the periphery of the spinning disk). The direction perpendicular to both these directions was from the center of the disk to its periphery, and it was indeed the direction in which the maximum voltage could be measured. The disk could produce direct current as it turned because the magnetic flux was continuously varying as new portions of the disk were carried into the magnet's gap. The virtue of this primitive dynamo was that the voltages induced between any two points on the surface could easily be assessed by sliding leads attached to a galvanometer. The defect of this generator is that the thick disk of highly conductive copper effectively short-circuits the induced voltages, so it is hard to extract sizable energy to the external galvanometer circuit. Most of the induced current resistively heats the copper disk.

One inspiration for this dynamo was the disk of Dominique Arago, which had been invented in 1825 but never explained [6]. Arago had noticed that a spinning copper disk created a magnetic field that would deflect a compass needle suspended over it. Faraday could now clearly explain that in a nonmagnetic copper disk spinning in earth's magnetic field, currents would be induced that could deflect a nearby compass needle.

Because it was the first advance in production of direct current beyond the improvements of Volta's battery, this invention was immediately recognized as both a scientific breakthrough and a very promising technology. When Prime Minister Robert Peel stopped by Faraday's lab for a demonstration, he asked what the new invention was good for. Faraday showed a good grasp of politics when he replied "I know not, but I'll wager your government will tax it" [2, p. 123]. And he was right, although a bit premature.

Having concentrated deeply on the problem on induction for several months, Faraday was pleased with the results as he published them, and he and Sarah escaped to Brighton for a well-deserved vacation. However, with results this breathtaking, it may not surprise you that another priority scandal brewed up. This one had its roots in a backdated Italian scientific journal. It was satisfactorily contained, but not without some wear and tear on the scrupulous Faraday. Faraday continued his electrical work for more than 20 more years but always subject to the distractions of the chemical analyses that were the major income stream of the Royal Institution.

Faraday loved his quiet basement laboratory, depicted in Figure 5.11, where he determinedly maintained his solitude when concentrating. He left standing orders on some days that he was not to be disturbed under any circumstances. But, of course, he was not really alone. He was having a deep and satisfying colloquy with Mother Nature. When he posed a question, she replied. And when the conversation got really interesting, well, those were the most treasured moments of his life.

This short summary cannot do justice to all of Faraday's important discoveries, but we must include what was his greatest contribution to theoretical physics: the theory of the field. Faraday's concept of a force field was the result of his deep and careful thinking on induction; it has grown to dominate theoretical physics ever since. Faraday knew what he could see, and iron filings sprinkled over a sheet of paper could detect and make visible the forces imposed on them by a nearby

Figure 5.11 Faraday in his basement laboratory at the Royal Institution. Painting by Harriet Moore.
Source: The Chemical Heritage Foundation collection.

magnet (Figure 5.12) or current-carrying wire. Even the electrostatic field around a charged object could be detected by small insulating particles. These forces can be seen to spring to life when the current flows or the magnet moves into position or the glass is rubbed, so they do not require the credibility straining the action-at-a-distance property hypothesized by Newton for the gravitational force.

Faraday published this work on induction and the magnetic field along with major papers on battery science as they were completed in the years 1832–1855. He included another groundbreaking discovery—what has since become known as *Faraday rotation.* After much experimenting, he showed that magnetic fields applied to some crystals can actually rotate the plane of polarization of light traversing them. He then published these collected papers with only minor modifications as *Experimental Researches in Electricity* in 1855. By this time he had learned his lessons about attribution, so he described the ideas of Ampére, with which he profoundly disagreed, as "Ampére's beautiful theory." This book is carefully written, although slow moving for our jaded modern tastes; it allows the patient reader to follow the thought processes of a great experimentalist. Mortimer Adler saw fit to include it in the series "Great Books of the Western World."

Fortified by his discovery of Faraday rotation, the scientist was able to make the leap of intuition that light is a vibration of the electromagnetic field. The mysterious *aether* that nineteenth-century optics had hypothesized as the medium that

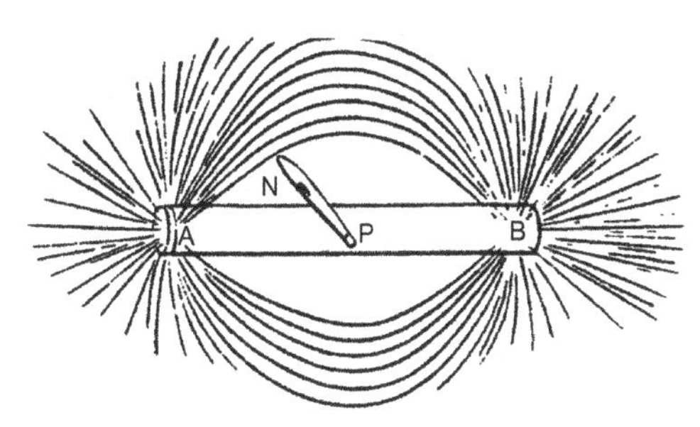

Figure 5.12 Lines of magnetic force around a magnet made visible by moving galvanometer probes.
Source: Faraday's *Experimental Researches in Electricity*. A and B are the poles of a bar magnet, and the silver knife blade NP was used to map out the field lines.

must fill space in order to propagate waves of light was not really necessary. The conjecture that light was an electromagnetic wave would be experimentally verified by Hertz in 1888. In modern quantum mechanics, the photon, or particle comprising light, is the elementary particle that communicates electromagnetic fields or forces.

Since Faraday was not able to mold his observations on induction and the electromagnetic field into mathematical form, that heroic task would be completed by James Clerk Maxwell in a series of papers starting in 1857 with "On Faraday's lines of Force." By 1861, Maxwell was able to calculate from the electrical and magnetic properties of empty space that electromagnetic waves should have a velocity very near to that measured for light. To most scientists, that evidence was convincing proof that light was an electromagnetic wave. Maxwell's contributions were not just in the nature of mathematical translation; he also added a new effect of *displacement current* not imagined by Faraday to make the equation set complete and internally consistent. Maxwell's equations now form the backbone of physics.

Faraday's intuition was not infallible. He had strongly criticized Ampére's conjecture that some materials manifest magnetism because they incorporate circulating charges that can align to produce magnetism, an idea rather close to our modern understanding that magnetic materials are made of magnetic atoms. Moreover, Faraday's enthusiasm for his field theory led him to hypothesize that centers of concentrations of lines of force were the fundamental constituents of matter. Faraday thus muddied the waters and ended up on the wrong side of the theory of the existence of atoms.

Like most people who triumph over poverty, Faraday believed in the virtue of economy. Like most great inductive thinkers, Faraday naturally consolidated superfluous hypotheses to gain simplicity. Therefore, it was an affront to him to believe that Nature had concocted two very similar central forces: electrostatic attraction and gravitation. It seemed to him to be one force too many. At some level, he believed, they must be closely related. Faraday attacked the problem of "gravelectricity" like Galileo on steroids: Among other experiments he had two 280-lb lead bricks repeatedly dropped from a tower in an attempt to measure a voltage induced on them that could account for their fall. He found no effect.

On the face of it, Faraday should have realized that his attack on Newton's gravitation was naive, because it would be hard to conceive a scheme that assigned unique charges to all of the planets and their moons that would rival the success of

Newton's gravitational theory. This struggle is a sad example of cutting too deeply with Ockham's razor. To quote Einstein, "Make things as simple as possible, but no simpler." Bundling the electrostatic force with gravity is simply untenable, as modern quantitative theory assesses that the electrostatic force is 10^{36} times as strong as gravitation, so they operate in different spheres.

Another simplification Faraday introduced was, however, crucial. In the third volume of his *Experimental Researches in Electricity*, he reports an exhaustive series of experiments comparing electricity obtained from electric fish, voltaic batteries, electrostatic generators and magnetic induction [6, Series III]. After carefully calibrating these sources so that he compared equal energies from each, he found that each affected a galvanometer similarly, heated a wire similarly, and had the same electrostatic repulsion and electrochemical effect. Faraday concluded that there is only one type of electricity. In these experiments, however, not every source was able to produce every effect. For instance, Faraday recognized that the electric torpedo fish just did not generate enough voltage to cause a spark. Thermoelectricity, the small voltage induced in a junction of dissimilar metals was likewise unable to generate all the effects. His final standard test, not surprising for one brought up in the era of the "electric fluid," was to note that if he removed the galvanometer and placed the two leads from the circuit on his tongue, the taste and sensation of all these electricities were qualitatively similar.

The wizardry Faraday showed in the laboratory is most reminiscent of that of Joseph Priestley. Faraday himself must have recognized some resemblance because he commented in 1833 at a meeting to mark the centenary of Priestley's birth, "Dr. Priestley made his great discoveries mainly in consequence of his having a mind which could easily be moved from what it had held to the reception of new thoughts and notions: and I will venture to say that all his discoveries followed from the facility with which he could leave a preconceived idea" [3, p. 265].

Faraday's contributions have been honored in many ways. His advances in understanding the nature of capacitance were so extensive that its unit has been named the *Farad*. Likewise, his wide-ranging and valuable electrochemical work has been honored by naming the unit of electrochemical charge the *Faraday*. He was awarded a share of the Copley Prize in 1832 "for his discovery of Magneto Electricity."

Faraday really never recovered reliable use of his memory after the attacks he suffered in the late 1830s. Nevertheless, he continued working intermittently into the 1860s. As a religious dissenter, he had early on accustomed himself to a place somewhat out of the mainstream of English society. Perhaps another reason why he seems to our modern sensibility to have remained strangely disconnected from the political sphere was that he was never a property owner and thus never able to vote. Ever the modest Sandemanian, or perhaps worried about his failing memory, he turned down a knighthood and two separate offers of the presidency of the Royal Society. However, in many of his projects, such as investigating the causes of major instances of mine pollution or explosions, or his campaign of 1855 to clean up the Thames River, he labored vigorously for the public good.

He gave his last lecture at the Royal Institute in 1862, after which he and Sarah retired to private life, maintained by a Civil List Pension and a house in Hampton

Court donated by Queen Victoria. He continued sinking into dementia and died in 1867. Perhaps it is significant that neither Faraday nor Davy had children; nineteenth-century chemical labs were toxic places.

We come away from Faraday's life story feeling that this was a man we would have loved to have known. He placed his life at the service of England's technological needs and his own genius, and he seldom wavered from his stony Victorian principles. All humanity benefited from his electrical discoveries, none of which he attempted to patent or exploit. The poor bookbinder's apprentice matured into the greatest experimentalist of his century.

References

[1] L. Pearce Williams, Michael Faraday, Basic Books, New York, NY, 1965, p. 26.
[2] A. Hirshfeld, The Electric Life of Michael Faraday, Walker & Co, New York, NY, 2006. p. 48.
[3] J. Hamilton, A Life of Discovery: Michael Faraday, Giant of the Scientific Revolution, Random House, New York, NY, 2002. p. 13.
[4] K. Jelved, Jackson A.D. Andrew, O. Knudsen, (Trans.) Selected Scientific Works of Hans Christian Ørsted, First American ed., Princeton University Press, Princeton, NJ, 1998, pp. 412–415.
[5] W.H. Cropper, Great Physicists: The Life and Times of Leading Physicists from Galileo to Hawking, Oxford, New York, NY, 2001, pp. 146–148.
[6] M. Faraday, Experimental researches in electricity, in: Hutchins (Ed.), Great Books of the Western World, Encyclopedia Britannica, Chicago, IL, 1952, Series I.

Bibliography

M. Faraday, Experimental researches in electricity, in: Hutchins (Ed.), Great Books of the Western World, Encyclopedia Britannica, Chicago, IL, 1984.
Truly a great book.
John, M. Gribbin, Faraday (1791–1867) in 90 Min, Constable, London, 1997.
M.J. Gutnik, Michael Faraday: Creative Scientist, Children's Press, Chicago, IL, 1986.
J. Hamilton, A Life of Discovery: Michael Faraday, Giant of the Scientific Revolution, Random House, New York, NY, 2002.
Written by a reader in the history of art. Readable, not technical, and emphasizes Faraday's artistic interests.
A. Hirshfeld, The Electric Life of Michael Faraday, Walker and Co., New York, NY, 2006.
Written by a physicist. Both lively and good technically.
F.A.J.L. James, Michael Faraday: A Very Short Introduction, Oxford University Press, Oxford, 2010.
Academic; not so interested in science.
L.P. Williams, Michael Faraday: A Biography, Clarion, New York, NY, 1965.
Exhaustive and well written.

6 Wilhelm Röntgen Intended to Study Cathode Rays but Ended Up Discovering X-Rays: Listen Carefully When Mother Nature Whispers in Your Ear—She May Be Leading You to a Nobel Prize

The discovery of X-rays by Wilhelm Röntgen is a clean example of the scientific method working beautifully. A well-prepared researcher in an adequately equipped laboratory devoted himself to verifying some new results in the exciting area of cathode rays. With just a little bit of luck, he observed a novel phenomenon unnoticed by previous researchers. By careful and thorough investigations, he documented the startling properties of X-rays in just 8 weeks. Only when he convincingly identified X-rays as electromagnetic radiation did he publish a paper and release the results to the scientific world. The popular press became entranced with the novelty of these rays and their bright promise and lionized him. In 1901, less than six years after his first publication, he received the inaugural Nobel Prize in physics as a triumphal recognition of his discovery.

The discovery was his alone: He had no help in the laboratory and kept his results secret from colleagues. No rival claimants appeared to demand a share of the discovery. Yet X-rays had been produced and overlooked without notice for decades in many laboratories of scientists studying gaseous discharges.

What sort of man could make this discovery? Röntgen was a journeyman physicist whose credentials were excellent, so he was well prepared to unravel the mystery of X-rays. Röntgen had been born in the German Rhineland in 1845 to a comfortably middle-class family. His father, Friedrich, was a merchant and manufacturer of cloth. His mother, Constanza, was a German with Dutch roots. In 1848, possibly in response to the political turmoil of that year, Friedrich moved the family to Apeldorn, Holland, and renounced his Prussian citizenship. Little Wilhelm was schooled in Apeldorn but was not an exceptional student, preferring to tinker with mechanical contrivances or spend his free time wandering in the out-of-doors. Both of these predilections remained with him for life. Even though a childhood

How the Great Scientists Reasoned. DOI: http://dx.doi.org/10.1016/B978-0-12-398498-2.00006-0

infection left him with only one good eye, he became a skilled and enthusiastic hunter in his adulthood.

When Wilhelm was 16 and a student at a secondary school in Utrecht, his steady progress through the educational system was suddenly blocked. As one of his teachers left the classroom, another student drew the teacher's caricature. The returning teacher became enraged and demanded that Wilhelm identify the culprit, but Wilhelm refused. Frustrated with his stubbornness, the teacher succeeded in getting Wilhelm expelled.

This seemingly minor incident began to loom very important in Wilhelm's life because he could not enter the university without a secondary school diploma. His family tried many methods of circumventing this problem: private study plus an exam, study at a university prep school, and entry as an unmatriculated student at the University of Utrecht. Only by escaping the Dutch system and enrolling at the Swiss Federal Institute of Technology (ETH) in Zurich was he finally able to surmount this difficulty and work toward a university *diplom*.

His years of study in Zurich were happy ones. He pursued his books diligently yet enjoyed the coffeehouses and taverns of Zurich with his classmates. He was known for being quiet and modest and was gently teased about his Dutch roots and accent. Though he made some lifelong friends, he was never fond of dancing or noisy parties. He met and fell in love with an attractive and amiable tavern keeper's daughter, Anna, 6 years his senior.

Under the guidance of a brilliant young experimental physicist, Professor August Kundt, Röntgen began the demanding but exhilarating experience of working in a research laboratory. After completing his diplom in mechanical engineering at the age of 23, he continued as a doctoral student under Kundt, completing his PhD with a thesis "Studies of Gases" at 24.

After graduation, he led the life of a peripatetic academic with rather catholic research interests: He followed Kundt to the University of Würzburg, from which he published papers on the capacity of gases to hold heat. He moved to Strasbourg in1872 where he began climbing the academic ladder as *privat docent*, felt secure enough to marry Anna, and published papers on a wide variety of topics—from a new type of barometer to the rotation of the plane of polarization of light in gases by electric fields (Kerr effect). A move to the University of Giessen netted him a professorship in 1879 and produced papers on the compressibility of gases and the surface tension and viscosity of liquids. He published a seminal paper from this period that clarified the forces on an insulator moving through an electric field.

Röntgen was developing a reputation as a classicist, an incisive researcher who exhaustively endeavored to solve problems completely, a gifted builder of scientific apparatus who was used to working alone, and a scientist with a great breadth of expertise and experience. He manifested Germanic thoroughness and discipline, priding himself on bringing every project to a successful conclusion. In 1888, after several offers, Röntgen accepted a position at the University of Würzburg as professor of physics and director of the Physical Institute. His photograph in Figure 6.1 shows a serious man with a penetrating gaze.

Figure 6.1 Wilhelm Röntgen.
Source: The Library of Congress.

Just how seriously devoted was Röntgen to the discipline of science? In a later lecture, Röntgen quoted as an ideal the words of the great engineer Werner von Siemens [1]:

> *The intellectual life gives us at times perhaps the purest and highest joy of which the human being is capable. If some phenomenon which has been shrouded in obscurity suddenly emerges into the light of knowledge, if the key to a long-sought mechanical combination has been found, if the missing link to a chain of thought has been fortuitously supplied, then this gives the discoverer the exultant feeling which comes with a victory of the mind, which alone can compensate him for all the struggle and effort, and which lifts him to a higher plane of existence.*

These words may sound very Victorian to us, but their unmistakable passion resonated in Röntgen.

And perhaps such passion was required at this time. During the last decades of the nineteenth century, a certain malaise hung over the field of physics: a feeling that most phenomena had been discovered and that perhaps the work of this and future generations would merely be devoted to the less exciting task of measuring physical constants to ever higher precision. Perhaps Röntgen himself coped with this discontent, but it would be his distinct honor to be one of the very first pioneers who would break through the wall of classical physics and illumine a quantum world that offered opportunities for a century of exciting discoveries.

Perhaps it will make Röntgen's work easier to understand if we first review X-rays and how they are produced [2]. X-rays are electromagnetic radiation, like light, but of much higher energy and frequency. A light quantum, depending on the exact color, has an energy of about 4 electron volts (eV), which means its energy is the same as that gained by an electron if it accelerates through a 4 volt potential difference. X-rays have energies of many hundreds or a few thousand of eV.

Beginning about 1870, physicists were beginning to take advantage of two inventions of Heinrich Geissler: the mercury column vacuum pump and glass tubes with sealed-in metallic electrodes (i.e., vacuum tubes). Used with the induction transformer, these inventions enabled scientists to study high-voltage discharges and observe the properties of beams of electrons that they called *cathode rays*. A wide variety of such tubes were built, but they shared the property that beams of electrons of energy above 1000 eV could be produced and directed toward the tube walls or metallic electrodes. Although it was completely unsuspected at the time, when high-energy electrons collide with surfaces, their sudden deceleration produces X-rays as shown in Figure 6.2.

Röntgen's first X-ray paper of January 1896 was an unusually complete description of a newly observed phenomenon [3]. It is written in a more discursive, chronological, and personal style than modern scientific papers, but it very nicely outlines the unique properties of X-rays and distinguishes them from cathode rays or visible or infrared radiation.

Röntgen began by describing his initial discovery: In November 1895, he had enclosed a Hittorf–Crookes tube with black cardboard so that no light originating from it could be seen. He did not specifically state this, but he seems to have intended to study any cathode rays that may have penetrated the glass walls. Working in a darkened room, he excited a discharge in the tube with an induction coil and noticed that a nearby cardboard sheet covered with the material barium platinocyanide fluoresced brightly while the discharge was maintained. Curiously, the fluorescence seemed as bright when the cardboard backing faced the tube as when the fluorescent coating faced the tube. With mounting excitement, he observed that the fluorescence was not attenuated when he interposed more layers of cardboard, two packs of cards, hard rubber several centimeters thick, and even a 1000-page book. Neither did 3 cm thick blocks of wood, water, or liquid carbon disulfide attenuate the fluorescence. Only when he interposed a 15 mm thick sheet of aluminum was the fluorescence "enfeebled considerably." Consumed with curiosity, Röntgen ate and slept in his lab for the next few days.

In fact, all of the metal sheets he tried attenuated the rays: copper, silver, gold, and platinum. Lead was the best attenuator, and even lead-containing glass (flint glass) attenuated the rays much more than normal glass. Röntgen really did not have an effective way to measure the attenuation, but denser metals seemed to be more effective attenuators. The attenuation was approximately proportional to the

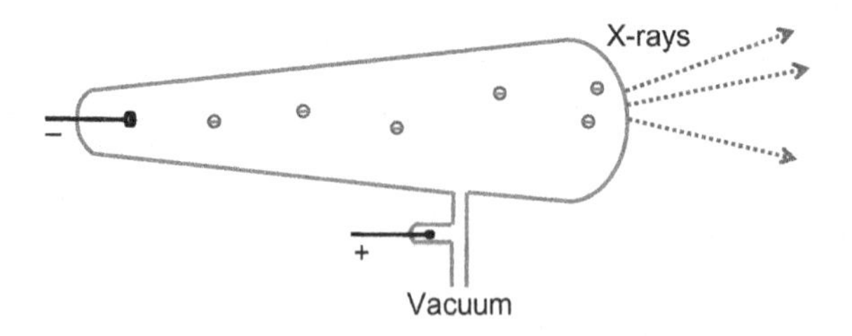

Figure 6.2 Röntgen's Hittorf–Crookes tube. Electrons (the circles with the minus signs) are ejected from the negative electrode and accelerated by the high voltage toward the positive electrode. Many overshoot to strike the glass envelope; their collisions excite X-rays (dashed lines).

metal's density times its thickness, but Röntgen cautioned that this observation was only a useful first approximation. He compared four materials of nearly equal density: glass, calcite, quartz and aluminum, and found that calcite was the least transparent of the group.

Turning his attention to the materials that might be used to detect the rays, he noted that not only barium platinocyanide fluoresced but also glasses of various types, calcium compounds, calcite, and rock salt to some degree. The rays also fogged photographic film, either on a glass or gelatin backing. Thorough to a fault, Röntgen cautioned that this fogging might be a multistep process, as the glass or gelatin might be fluorescing inside the cardboard-wrapped film plate, so it was not entirely certain that the film itself was detecting the rays or secondary light. Furthermore, his eye was not sensitive to the rays: Staring directly at the source of the beam did not excite the retina. These observations pushed Röntgen to think more deeply about what energy the rays might be transferring, and he noted that the rays probably heat materials they pass through, but he was unable to measure any heating.

So these rays act somewhat like light, but they are obviously much more penetrating than light. Röntgen tried to observe some optical behavior. Unlike light, he found that the rays were not measurably deviated by prisms of glass, water, or carbon disulfide, calculating that that the coefficient of refraction for the rays in these materials must be smaller than 1.05—not much different from air. Röntgen thought that small crystallites might act like mirrors or lenses to reflect or concentrate the rays, but he identified no scattering by powdered rock salt, silver, or zinc that was any different from the bulk materials. Large lenses of glass or hard rubber did not concentrate the rays; the rays were simply attenuated more in traversing the thicker portions of both lenses.

Removing the cardboard shield from the Crookes tube, Röntgen saw that the rays originated from a glowing spot on the wall of the glass tube.

Reinstalling the cardboard shield, Röntgen could affirm that the rays were attenuated according to the inverse square law. That meant that their intensity was only one-quarter as great 200 mm from the glowing spot on the tube wall as it was at 100 mm. This would resemble the behavior of light waves radiating from the glowing spot. This observation proved that the rays are not cathode rays, as did his observation that the rays were not deviated by a magnet.

Röntgen now hazarded his major conclusion: "[The] X-rays are not cathode rays, but ... are produced by the cathode rays at the glass walls of the discharge apparatus." Furthermore, in a tube with a 2-mm-thick aluminum wall, he verified that the X-rays originate from the aluminum.

He further justified the term *rays* by citing photographs of a compass in which the magnetic needle was clearly visible, although it was in a metal case, and a metal sheet showing some interior heterogeneities. Perhaps the most stunning part of the paper is a reference to a photograph of the bones and ring of Anna's hand (Figure 6.3).

Röntgen maintained, but did not show, that he made an identifiable but weak pinhole photograph of the cardboard-clad X-ray apparatus. His efforts to detect

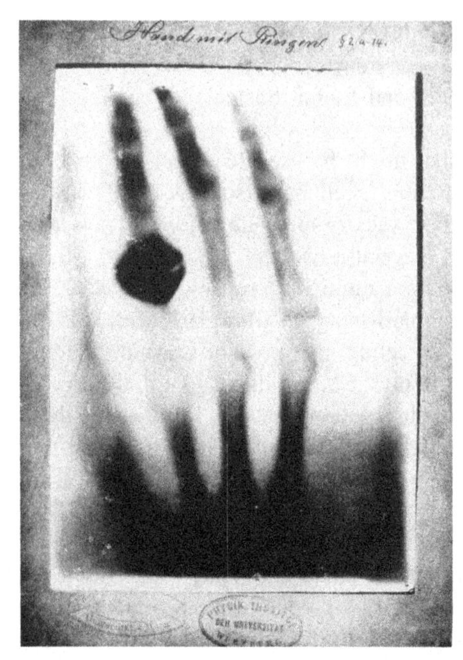

Figure 6.3 Röntgen's original photo of Anna's hand and rings, 1895.
Source: The Deutsches Museum.

interference phenomena were unsuccessful. (This is not surprising: The X-rays were an incoherent mixture of different short wavelengths.)

Clearly, Röntgen states, these new rays are not ultraviolet light, as they are not refracted or reflected and cannot be polarized by "ordinary methods."

Röntgen now concluded that these new rays are very much like light in that they cast shadows and cause chemical action such as fluorescence. He used the nineteenth-century designation "vibrations of the ether" rather than using the modern designation *electromagnetic waves.* Casting about for some explanation of how the rays can be like light but so palpably different [2, pp. 821–828], he took a misstep by speculating that the rays were longitudinal rather than transversely polarized waves as light is, reasoning that sound waves can be longitudinally polarized, so why not light? But Maxwell's electromagnetic theory, complete and gaining acceptance at that time, had shown that longitudinally polarized electromagnetic waves should be strongly damped.

The publication of Röntgen's paper "On a New Kind of Rays" in January 1896 with the contents we have just summarized was a remarkable scientific achievement. It described the novel and useful properties of a new type of radiation, clearly distinguished it from previously known rays, and identified it as an electromagnetic ray related to light. He discovered its most useful property: its ability to peer beneath living tissue. His achievement is underscored when we realize that the apparatus he used to produce and investigate these rays had been in service in many laboratories in many countries for more than 20 years without their users observing X-rays.

Therefore, in early 1896, as Röntgen's discovery was being evaluated and confirmed in other laboratories, many notable scientists realized that they had been scooped. For some of the greatest scientists of the nineteenth century, January 1896 must have been the "Ah shit!" moment of a lifetime. To their credit, most generously praised Röntgen's work and lauded his discovery. A. Swinton in England began circulating X-ray photographs of a hand "according to Professor Röntgen's method," while J.J. Thompson at Cambridge began his own experiments, as did other pioneers in France, Canada, and the United States [1, pp. 33–36]. In mid-January, Röntgen gave a demonstration to Kaiser Wilhelm and other notables in Berlin, answering questions until late in the evening and receiving the Prussian Order of the Crown, Second Class. Figure 6.4 illustrates how the new rays were immediately celebrated in the popular press. Nevertheless, a few sore losers took refuge in the belief that Röntgen had just been "lucky."

The fact is that a scientist working alone in a laboratory is engaging in a bare-knuckles brawl with a secretive Mother Nature. Very often, one phenomenon that

Figure 6.4 *Life* magazine's take on X-ray photography, February 1896.
Source: Jeremy Norman & Co.

an apparatus has been designed to measure is overpowered or masked by another. Moreover, the continuously increasing entropy of the natural world scatters errors carelessly through even well-planned experiments, and following up each false lead can be exhausting. Even so, scientists must constantly strive to keep their minds open to the novel and unplanned. This may be particularly difficult for them because, frequently by temperament and always by training, they are intellectuals who long to impose meaningful order on the sometimes chaotic world of nature.

Here are just two of the many missed opportunities to discover X-rays: A.W. Goodspeed and W.N. Jennings at the University of Pennsylvania were photographing electric sparks in 1890 [1, pp. 222–223]. Their workbench became littered with photographic plates, tools, and equipment. Jennings developed some of the plates at the conclusion of the day's experiments and observed shadowgraphs of the equipment, but he did not pursue further experiments. Doubtless these experimenters had too many important things to take care of in their laboratory to follow up on what may have simply been a poorly manufactured batch of photographic plates, but you may be certain they were suffused with remorse when they later read of Röntgen's work. In another similar irony, William Crookes revealed that he had angrily returned several batches of photographic plates stored near his eponymous tubes to the manufacturer, complaining that he had received them in fogged condition [1, pp. 223–224].

The lesson of mysteriously fogged photographic plates had been well assimilated by the scientific community when, later in 1896, Parisian physicist Henri Becquerel noticed that photographic plates in contact with a uranium salt were fogged even before they were placed in sunlight to excite the phosphorescence he intended to study [4]. For his subsequent discovery of natural radioactivity, Becquerel was awarded the 1903 Nobel Prize in physics, sharing it with Pierre and Marie Curie.

Is there anything that Röntgen overlooked in that 8-week period of intense research that might have added to his accomplishment's luster? The discovery by William H. and William L. Bragg, father and son, of the diffraction of X-rays by crystals, for which they were awarded the 1915 Nobel Prize probably demanded a more intense and monochromatic (i.e., better-tuned) X-rays source than what was available to Röntgen [2]. But it is certainly to Röntgen's credit that he tried several experiments with crystals and crushed crystallites which could have revealed X-ray diffraction had his source been adequate. Röntgen deserves some credit for having good intuition since he evidently had caught the scent of an important discovery.

Perhaps Röntgen benefited from refocusing on the completely new field at a period when his powers were at their peak. He was unburdened with the stale conclusions of those who had worked with cathode rays for many years. He could turn fresh eyes to careful reexamination of its novelties.

In short, Wilhelm Röntgen went into the laboratory to study cathode rays, got a secretive little kiss from Mother Nature, and kept his eyes and mind open enough to make one of the great discoveries of modern physics. As we mentioned, he was honored with the inaugural Nobel Prize in 1901. The unit of ionizing (high energy) radiation was named after him in the 1920s, and element 111 was named Roentgenium by its discoverers in Darmstadt, Germany, in 1994.

You might be wondering how the early pioneers of X-ray technology protected themselves from the radiation they studied. Not at all, of course. Like Michael Faraday, Sir Humphry Davy, and Lavoisier, Röntgen had no children. Nevertheless, he lived a long and active life, surviving the difficult periods of World War I and the great postwar inflation in Germany. He outlived his frequently ill Anna to die in 1923 of intestinal cancer at the age of 79.

References

[1] O. Glasser, Wilhelm Conrad Röntgen and the Early History of Röntgen Rays, Charles C. Thomas, Springfield, IL, 1934, p. 75.
[2] F.W. Sears, M.W. Zemansky, H.D. Young, University Physics, seventh ed., Addison-Wesley, Reading, MA, 1987, pp. 938–939.
[3] W.C. Röntgen, On a new kind of rays, in: O. Glasser (Ed.), Wilhelm Conrad Röntgen and the Early History of Röntgen Rays, Charles C. Thomas, Springfield, IL, 1934, pp. 16–28.
[4] W.H. Cropper, Great Physicists: The Life and Times of Leading Physicists from Galileo to Hawking, Oxford University Press, New York, NY, 2001, p. 299.

Bibliography

O. Glasser, Wilhelm Conrad Roentgen and the early history of Roentgen Rays, Charles C. Thomas, Springfield, IL, 1934.
R.R. Nitske, The Life of Wilhelm Conrad Röntgen, Discoverer of the X Ray, University of Arizona Press, Tucson, AZ, 1971.

7 Max Planck, the First Superhero of Quantum Theory, Saves the Universe from the Ultraviolet Catastrophe: Assemble Two Flawed Hypotheses About a Key Phenomenon into a Model That Fits Experiment Exactly and People Will Listen to You Even if You Must Revolutionize Physics

Before the turn of the twentieth century, a few severe cracks were becoming apparent in the edifice of basic physics. The clearest example of these was that physicists' best understanding of the radiation emitted by hot bodies was very, very wrong.

It is a familiar observation that if you take a hot poker out of a roaring fire, you can infer its temperature from its color. As a very hot poker cools, its color will change from a bluish white at about 1100°C to a dull red at 800°C. However, it is not simple to exactly measure the temperature of an incandescent body from its radiated light because some materials have surfaces that emit more radiation than others at the same temperature.

But physics proceeds by searching for absolutes, and it occurred to Gustav Kirchhoff in 1859 that one could circumvent this problem by studying only radiation at equilibrium with hot bodies. Each area of the interior of a closed furnace at constant temperature absorbs as much radiation as it emits because its temperature remains constant. Therefore, the electromagnetic radiation, ranging from the lowest energy infrared, through the visible, and into the most energetic portion of the ultraviolet spectrum in equilibrium with the hot walls of any such cavity should be identical at identical temperatures. This proposition became known as Kirchhoff's law [1]. One should expect to be able to calculate the salient features of such radiation at any temperature and compare it with experiments

How the Great Scientists Reasoned. DOI: http://dx.doi.org/10.1016/B978-0-12-398498-2.00007-2

because this *cavity radiation* can be examined in the laboratory by cutting a hole of negligible size in a furnace and analyzing the emitted light. Cavity radiation is sometimes called *black-body radiation* because black surfaces emit radiation so efficiently that they approximate the radiation coming through a small hole in a furnace. Figure 7.1 compares the spectra of cavity radiation for two different temperatures.

Since the characteristics of this radiation should be an absolute of physics, it was a great disappointment that when Lord Rayleigh [2], one of the top guns of mathematical physics in 1900, published a theory of the spectrum of cavity radiation, his expression differed markedly from measurements.

The Rayleigh theory starts with notions such as the theory of the resonant wavelength of a vibrating string that you learned in high school physics. Because the ends of the string are clamped in position, they cannot move, even though the rest of the string is free to vibrate. The vibrations of the string can last a long time, or *resonate*, only when the length of the string is an integral number of half-wavelengths of the sound wave long because the movement of the string is always zero at the ends and half-wavelength points. So a vibrating string of length L can support resonant waves of wavelength [3]

$$\lambda = \frac{2L}{1}, \frac{2L}{2}, \frac{2L}{3}, \frac{2L}{4}, \ldots \tag{7.1}$$

resembling those of Figure 7.2, which plots displacement versus distance along the axis of the string. Each resonant wave is said to be a *mode of vibration* of the string. We have plotted the five longest wavelength resonant waves that satisfy the condition that the ends of the string are clamped in position.

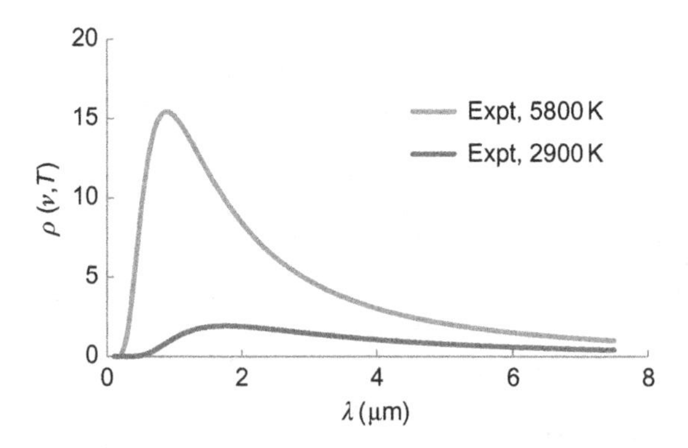

Figure 7.1 Spectra of cavity radiation for two different temperatures. A temperature of 5800 K is chosen because it is the temperature of the sun's surface. $\rho\ (v, T)$ is the total energy stored in a unit volume of a cavity at temperature T per interval of frequency (in units of 10^{-16} J s/m^3). Wavelength is expressed in micrometers (1 µm = 10^{-6} m). The wavelength of the 5800 K peak corresponds to visible light. Wavelengths longer than this correspond to the infrared. Wavelengths shorter than this correspond to the ultraviolet.

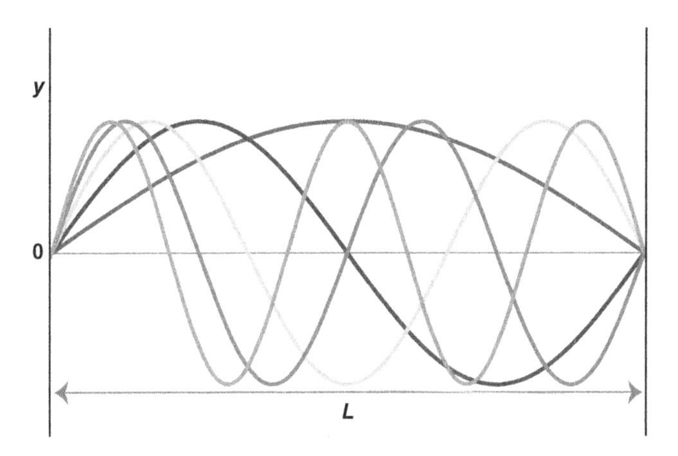

Figure 7.2 Some resonant wavelengths of a string of length L. The maximum displacement of each resonant wave (y) is plotted along the length of the string, with the ends being clamped to the wall on the right and left.

Electromagnetic waves such as heat radiation confined within a cavity with electrically conductive walls should also die away at the boundaries because they induce electric currents that dissipate their energies at the walls. For radiation confined between two planes a distance L apart, only wavelengths long enough to fit an integral number of half-waves into the cavity can resonate. For such waves, the number of half-wavelengths n for each wave of length λ resonating within the cavity will be, just as for the vibrating string,

$$n = \frac{2L}{\lambda}. \tag{7.2}$$

Note that for short wavelengths of radiation, there are going to be very many possible resonant waves.

It is frequently convenient to express these modes in terms of frequency rather than wavelength. Figure 7.3 shows how the product of wavelength λ and frequency ν is just the velocity of light, just as the product of the length of each boxcar on a long freight train times the number of boxcars passing per second equals the train velocity.

Our expression for the number of half-wavelengths resonant in a one-dimensional cavity now can be expressed in terms of the light's frequency ν and becomes

$$n = \frac{2L\nu}{c}. \tag{7.3}$$

Lord Rayleigh tried to calculate how much electromagnetic energy could be stored in such a cavity. In principle, this should be just the total number of resonant modes possible times the energy stored in each mode. Summing up all the resonant

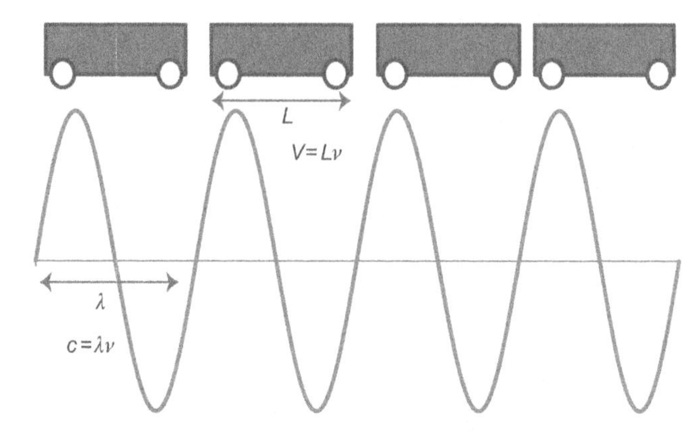

Figure 7.3 Schematic illustrating how frequency and wavelength are related. The velocity of a train whose cars pass ν times a second and are L long will be νL. The velocity c of a light wave of frequency ν and wavelength λ is $\nu\lambda$.

modes in a three-dimensional cavity is a tedious physics problem, and Rayleigh [4] took 5 years to derive a result incorrect by a factor of 8. Rayleigh's error was soon corrected by James Jeans, but we will not even attempt to derive it here [5]. Qualitatively, however, the number of possible resonant waves with a frequency less than ν in a three-dimensional cavity is related to the cube of the number of one-dimensional modes possible. Therefore, the number of modes is proportional to ν^3. And how much energy is stored in each resonant mode? Well, each heat or light wave must initially get its energy from a hot, vibrating atom of the cavity surface. The energies should be similar to those of a molecule in a gas, which Ludwig Boltzmann had shown more than 30 years earlier to be proportional to the absolute temperature T (using the Kelvin scale, which measures from absolute 0 at 273°C below the freezing point of water) [6].

Rayleigh's first crack at the problem then correctly deduced that the total energy per unit volume that would be stored in a cavity in equilibrium at temperature T *per frequency interval* ν would be

$$\rho(\nu, T) = C_0 \frac{T\nu^3}{\nu} = C_0 T \nu^2, \tag{7.4}$$

where C_0 is a constant to be determined.

In Figure 7.4, we plot the completed and correctly normalized Rayleigh–Jeans law,

$$\rho(\nu, T) = \frac{8\pi R T \nu^2}{c^3 N}, \tag{7.5}$$

with the constant C_0 evaluated by both Einstein and Jeans independently. The formula uses the speed of light c, the gas constant R, and the number of molecules per

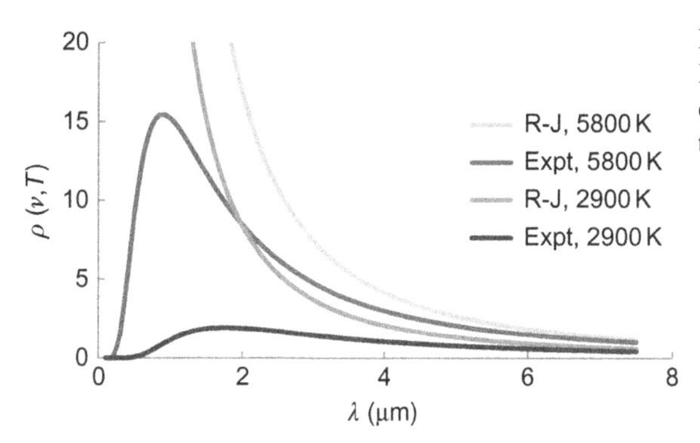

Figure 7.4 Comparison of the Rayleigh–Jeans theory to experiment at two different temperatures.

mole N (Chapter 8 will describe these constants more fully). Figure 7.4 compares the Rayleigh–Jeans law with the best experimental data for two different temperatures. They are a good approximation to measurements at large wavelengths, but for short wavelengths (the left-hand side of the plot), they depict the *ultraviolet catastrophe*: The Rayleigh–Jeans–Einstein theory predicts that the amount of energy stored in the cavity at short wavelengths approaches infinity.

The man who unraveled this dilemma did it by taking the first small step toward the development of quantum mechanics. Max Planck was born in Kiel, in the Danish province of Holstein in 1858, the sixth child of a law professor and the grandson of theologians on both sides of his family. Three years after Holstein was annexed by the Prussians in 1864, the Planck family moved to Munich, where Max received both his secondary and university educations. Though he showed promising musical gifts, Max was strongly attracted to science because it promised what he yearned for—absolute truth. Planck remembered vividly how his first physics teacher illustrated the law of conservation of energy by describing a mason laboriously carrying a stone up a ladder to cement it into the top of the wall. Centuries later, the stone's mortar crumbled, allowing the stone to fall to Earth with a crash; its potential energy had been stored unaltered for lifetimes. Max realized that the unalterable and universal truth he sought could be best found in physics. His photograph from that period (Figure 7.5) shows a very academic-looking Planck peering at the world through his pince-nez.

Planck centered his career steadfastly on the most conservative and trustworthy portion of physics' thermodynamics. He writes so modestly about his career in his *Scientific Autobiography* that one sees it as a lackluster climb up the German academic ladder from Munich to Berlin to Kiel, and finally to a full professorship in Berlin. He attributed any successes to luck or family influence. His field was always thermodynamics, and specifically entropy, where some of his best work was presaged by American Josiah Gibbs, leaving Planck with little credit. Entropy is a somewhat more exotic creature than energy but almost as important. Applying the law of conservation of energy to systems more complicated than a stone sitting on a wall frequently must take into consideration changes in entropy, or thermodynamic disorder. For instance, adding an ice cube to a glass of water allows

Figure 7.5 Planck in 1878.

crystalline ice to increase its entropy by melting; it must extract the required energy of melting by cooling the glass of water. For molecules of water, entropy increases as they go from ice to liquid to vapor. In general, every process involving melting or boiling or change in phase or chemical state will be accompanied by an increase in entropy. Planck did important work in clarifying the concept of entropy in chemical equilibrium, but at the time of his work on cavity radiation, his name was far from a household word in Germany. Still, Planck recognized before 1900 that the mystery of cavity radiation was the type of problem he yearned for because this radiation was an absolute that was independent of any material, yet incompletely understood.

In the summer of 1900, Planck was not very impressed with Rayleigh's first version of his radiation law (Eq. (7.4)) because it presumed that the radiation energy was equally distributed among all of possible resonant modes, and that was the reason it falsely predicted an ultraviolet catastrophe. Planck had derived an expression similar to Eq. (7.4) himself, but his version did not explicitly energize all the high-frequency modes and go to infinity at high frequencies. At the time, Planck had been trying to derive a theoretical justification of an equation that was known to successfully fit the density of cavity radiation at temperature T per frequency interval ν at small wavelengths,

$$\rho(\nu, T) = \frac{c_1 \nu^3}{c^3} e^{-\frac{c_2 \nu}{kT}}. \tag{7.6}$$

This equation had been proposed by his colleague Wilhelm Wien in Berlin in 1896 [1, p. 44]. The Wien radiation law avoids the ultraviolet catastrophe because as ν becomes large, the exponential term becomes small even faster than ν^3 becomes large, so it keeps $\rho(\nu,T)$ from becoming infinite. Wien's law had not had a compelling theoretical justification, and the constants c_1 and c_2 simply had to be determined by fitting experiment. Planck's attempt to theoretically justify Wien's law was based on selecting the most promising expression that maximized the entropy of the (not explicitly described) resonators emitting and absorbing cavity radiation.

He expected his procedure to represent cavity radiation's equilibrium distribution and lead to Eq. (7.6).

However, in October 1900, two experimental groups in Berlin reported new measurements of $\rho(\nu, T)$ probing farther into the long-wavelength region of the infrared and showing definite disagreement with Wien's law (Figure 7.6). In fact, at long wavelengths, the results more closely resembled the Rayleigh law. Planck had been barking up the wrong tree!

Planck [7] set to work immediately to come up with a radiation law that could describe cavity radiation in both regimes. Planck's synthesis of these two laws,

$$\rho(\nu, T) = \frac{c_1 \nu^3}{c^3} \frac{1}{e^{(c_2 \nu / kT)} - 1}, \tag{7.7}$$

built a continuous, smooth bridge between them. This expression fit the new data to within experimental accuracy and improved measurements coming from later experimental groups just coincided with this Planck radiation law more exactly. Planck's equation was at first just an inspired bit of curve fitting, but nevertheless it had been guided by his search for the proper theoretical derivation for Wien's law. At first, it was unclear why this equation worked so well, but Planck was fascinated by it and used it to launch his theoretical attempts to discover the underlying reason why it described the spectrum of cavity radiation so much better than the current theory of radiation. The equation avoided the ultraviolet catastrophe and somehow framed the key questions: Why should electromagnetic waves of short wavelength or high frequency be excited so sparingly within the cavity? Could there conceivably be a reason why the thermal energy from the cavity wall could not excite the high-frequency waves?

Planck devised a theoretical justification for this law during the next 2 months with a calculation appreciatively described by Abraham Pais, the dean of physics biographers, as "a little bit mad [8]." Planck began by using standard thermodynamics to evaluate the entropy associated with cavity radiation described by his radiation law, Eq. (7.7). His next step was completely unconventional and unprecedented: He calculated the entropy of a large collection of resonators that

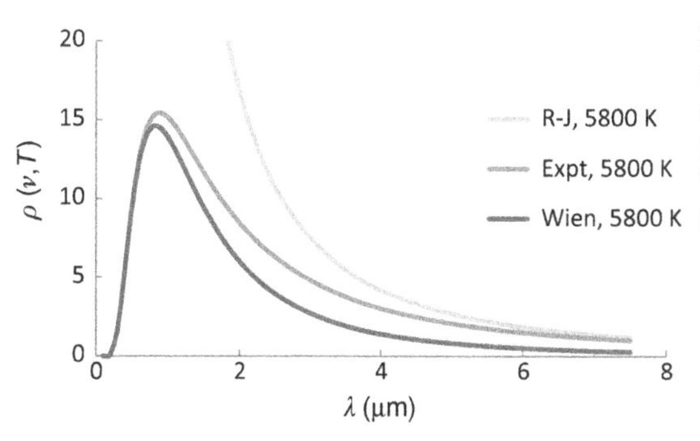

Figure 7.6 Planck radiation law (coinciding exactly with experiment) compared to Wien's law and the Rayleigh–Jeans law. The Planck radiation law is identical to experiment.

represented the cavity radiators by apportioning small "quanta" of discrete energy ϵ among them and determining their entropy. These expressions were identical if

$$\epsilon = c_2\nu = h\nu. \tag{7.8}$$

Planck [9] here renamed one of his two unknown constants in his radiation law h and was able to evaluate the other, with this derivation. It yielded the final form

$$\rho(\nu, T) = \frac{8\pi h\nu^3}{c^3}\frac{1}{e^{(h\nu/kT)} - 1}. \tag{7.9}$$

This gave the radiation's energy density in the cavity $\rho(\nu,T)$ at equilibrium at temperature T in a frequency interval around ν. The newly derived formula incorporated the velocity of light ($c = 3 \times 10^8$ m/s) and the Boltzmann constant ($k = 1.38 \times 10^{-23}$ J/K). What about the remaining constant h? This constant has dimensions of (energy × time), which physicists know as *action*, and Planck evaluated it by comparison with experiment to be $6.55 \cdot 10^{-34}$ J s, close to the modern value of $6.63 \cdot 10^{-34}$ J s. He was completely unable to relate this constant to any previously known properties of matter or radiation. For the frequencies he was considering, the energies it described, $h\nu$, were very small numbers.

Planck later wrote of his difficulties in accepting that his constant h, which he had christened "the elementary quantum of action," was really something so novel that it lay outside the realm of classical physics. "My futile attempts to fit the elementary quantum of action somehow into the classical theory continued for a number of years and they cost me a great deal of effort" [10].

This notion that the energy of these resonators must be *quantized* and could not assume a continuous range of values represented the birth of quantum mechanics. With the energies of higher-frequency modes rising with frequency as $h\nu$, Planck's law explained why the ultraviolet catastrophe could not actually happen; modes of high frequency require large amounts of energy to excite. Thermal excitation of the resonators vibrating at very high frequencies is just not energetically possible, so electromagnetic waves of these frequencies cannot be produced in large numbers, even though their frequency may fit the condition for resonance

$$\nu = \frac{nc}{2L} \tag{7.10}$$

with n being an integer, c is the velocity of light, and L is the distance across the cavity.

Ironically, Planck had had a long-running feud with Boltzmann about two issues: Were atoms real? Could the classical prohibition against entropy decrease be replaced by Boltzmann's revolutionary hypothesis that entropy decrease is not in principle strictly forbidden but only statistically unlikely? It was a measure of Planck's greatness that, although he had conservatively (and incorrectly) arrayed himself against both of these proposals for years, he was able to embrace both the reality of atoms and the statistical nature of entropy when he saw that these principles were required for his solution to the riddle of cavity radiation. In a further irony, Planck even succeeded in breathing clarity into Boltzmann's own theory by explicitly defining the quantity k as

$$k = \frac{R}{N}. \tag{7.11}$$

Perhaps somewhat unfairly, k is now known as the *Boltzmann constant* and is proportional to the average energy stored in a gas molecule by virtue of its confinement at temperature T. Boltzmann had used the gas constant R divided by N, the number of molecules in a mole, somewhat obscuring the molecular nature of this energy. Planck had even been able to determine N by comparing his radiation law with experiment, giving 6.175×10^{23} molecules per gram molecule, compared to its current value of 6.02×10^{23} (see Chapter 8 for a complete exposition).

Later in life, Planck ruefully mused in his *Scientific Autobiography* on the many years it took for Boltzmann's ideas to gain acceptance: "A new scientific truth does not triumph by convincing its opponents and making them see the light, but rather because its opponents eventually die, and a new generation grows up that is familiar with it" [10, p. 82]. Planck even anticipated Thomas Kuhn!

Planck's successful struggle with these questions marked the instant when quantum mechanics was born. Although his derivation was too unconventional for it to be rapidly accepted, it provided Albert Einstein the inspiration to explain the photoelectric effect and Niels Bohr the rationalization for the first explanation of the atom. Planck received the Nobel Prize for this work in 1918; his Nobel photographic portrait is shown in Figure 7.7.

Ironically, even though Einstein built his Nobel Prize-winning explanation of the photoelectric effect on Planck's idea of quantized energy, Planck thought that Einstein had stretched too far in quantizing light. However, the two became fast friends, and their musical evenings together in Berlin with Planck at the piano and Einstein on the violin were events both men looked forward to.

Planck remained a conservative throughout his life, and his *Scientific Autobiography*'s spirited defense of religion, brushing aside the miraculous and

Figure 7.7 Max Planck in 1919.
Source: A.B. Lagrelius and Westphal.

affirming religion's enduring humane and ethical components, is a classic. Planck had the luck to be raised from adolescence in Bavaria, where a jolly approach to enjoying life is celebrated, while reaching the peak of his career in Berlin, where the more serious and professional virtues are cultivated. His lectures were accomplished and clear, though not personable.

He was strong and physically active to the end of his life. Visitors to his Berlin home were likely to be drafted into a vigorous game of tag in his garden, where Planck was agile and elusive. Planck loved hiking in the mountains with such stamina and concentration that his outings were considered to be forced marches.

Planck suffered deeply as the disastrous history of Germany in the twentieth century unfolded. His first wife, Marie, died in 1909. His second son was killed in the First World War at Verdun. One of Planck's daughters died giving birth to his granddaughter; her twin married the widower and died in childbirth herself 2 years later. Planck did not recklessly defy the Nazis during the Third Reich, but he made himself sufficiently unpopular with them by continuing to teach Einstein's theories and supporting Jewish scientists that he lost his prestigious position as head of the Kaiser Wilhelm Gesellschaft. His eldest son, who had been captured by the French in 1914, lived to take part in the failed Valhalla plot of 1944 to assassinate Hitler and was executed by the Nazis in 1945. Planck's home in the Berlin suburbs was flattened by a bomb in 1944, with the loss of his personal papers. Planck died in 1947 with only one of his five children, the son of second wife Marga surviving him.

Germany has honored Max Planck by dedicating its prestigious chain of nearly 80 basic research institutes in his name. His life was recently celebrated with the minting of a commemorative coin in 2008 (Figure 7.8). He embodied the finest productive virtues that we associate with Teutonic culture; rectitude, laborious, and careful craftsmanship, and a thoroughgoing sense of responsibility.

Figure 7.8 A 10-euro coin minted in 2008 commemorating Planck's radiation law and 150th birthday.

References

[1] J.R. Mahan, Radiation Heat Transfer: A Statistical Approach, Wiley, New York, NY, 2002, pp. 80–81.
[2] J.W.S. Rayleigh, Remarks upon the Law of Complete Radiation, *Phil. Mag.* 49 (1900) 539–540.
[3] R.L. Lehrman, Physics the Easy Way, Barron's, New York, NY, 1990, p. 177.
[4] J.W.S. Rayleigh, The dynamical theory of gases and radiation, Nature 72 (1905) 54–55.
[5] J.H. Jeans, On the partition of energy between matter and aether, Phil. Mag. 10 (1905) 91–98.
[6] W.H. Cropper, Great Physicists: The Life and Times of Leading Physicists from Galileo to Hawking, Oxford University Press, New York, NY, 2001, pp. 179–199.
[7] M. Planck, On an improvement of Wien's equation for the spectrum, in: D. ter Haar (Ed.), The Old Quantum Theory, Pergamon, London, 1967, pp. 79–81.
[8] A. Pais, "Subtle is the Lord...": The Science and the Life of Albert Einstein, Clarendon Press, Oxford, 1982, p. 371.
[9] M. Planck, On the law of the distribution of energy in the normal spectrum, in: D. ter Haar (Ed.), The Old Quantum Theory, Pergamon, London, 1967, pp. 83–105.
[10] M. Planck, Scientific autobiography, in: R.M. Hutchins (Ed.), Great Books of the Western World, Encyclopedia Britannica, Chicago, IL, 1952. p. 84, Series I.

Bibliography

W.H. Cropper, Great Physicists: The Life and Times of Leading Physicists from Galileo to Hawking, Oxford University Press, New York, NY, 2001.
M. Planck, Scientific autobiography, in: R.M. Hutchins (Ed.), Great Books of the Western World, Encyclopedia Britannica, Chicago, IL, 1952, Series I.
D. ter Haar (Ed.), The Old Quantum Theory, Pergamon, London, 1967.

8 Albert Einstein Attacks the Problem "Are Atoms Real?" from Every Angle: Solving a Centuries-Old Riddle in Seven Different Ways Can Finally Resolve It

The man who was to make unparalleled contributions to twentieth century physics, the man whose very name became synonymous with genius, was born in Ulm, Germany, in 1879. Albert Einstein's family was secular Jewish on both sides. His father, Hermann, was known for his genial disposition and impractical business decisions. Young Albert reportedly worried his parents by being slow to learn to speak, but he soon became an excellent scholar and was able to enter a demanding gymnasium in Munich. When his family's electrical contracting business met reverses in Bavaria and moved to Milan, they left Albert behind to complete his studies. Albert coped with the loneliness for awhile, but he could not endure the harsh and arbitrary authority of the German gymnasium, and he both withdrew from school and determined to surrender his German citizenship to avoid the draft. After an extended visit to his family in Italy, Albert was able to finish preparatory school at a far more liberal Swiss school in Aarau. He later recalled, "The school left an indelible impression on me because of its liberal spirit and the unaffected thoughtfulness of its teachers, who in no way relied on external authority" [1].

Albert fell in love with Switzerland and on his second attempt passed the entrance exam for Zürich Polytechnic Institute, determined to prepare himself to be a secondary school teacher in a gymnasium. Hermann's business was doing poorly, so Albert had to content himself with a meager allowance, yet he pronounced himself content with "strenuous labor and the contemplation of God's nature..." as he began his university studies [1, p. 41]. He found he loved physics, but chafed as always under the discipline and authority of his professors. By his last years at the Poly, he became so disaffected that he spent little time in classes and much time reading physics texts in his room. After a traumatic bout of cramming, Einstein did pass his finals, leaving behind an outstanding examination record but horrible laboratory grades and poor recommendations from the Poly faculty. Consequently, Einstein could not find a permanent teaching position after graduation, and the

How the Great Scientists Reasoned. DOI: http://dx.doi.org/10.1016/B978-0-12-398498-2.00008-4

following 2 years were trying ones for him as he scrambled to support himself with tutoring and substitute teaching. Even so, this stormy education did not permanently diminish Einstein's love of physics.

During this period of marginal employment, Einstein deepened his affair with classmate Mileva Marić (Figure 8.1). Their love letters are touching and sometimes passionate; he had pet names for her like "Dollie" and "you little Witch," and she called him "Johnnie" [2]. Einstein had to secure a permanent job before he could consider marriage, and the best opportunity available was an unprestigious position as technical expert third class at the Bern patent office. When Mileva's unplanned pregnancy in 1902 presented a possible barrier to Einstein's entering the stolid Swiss civil service, she discreetly returned to her native Novi Sad, Hungary, to bear their daughter, Lieserl. She found someone to care for the infant and returned to Bern as soon as possible. Einstein and Marić effectively erased Lieserl from the historical record after this point.

Albert and Mileva married in early 1903, about a half-year after Einstein began his labors at the patent office, and settled down in Bern to enjoy what wedded bliss was allotted to them. The stimulating yet undemanding work at the patent office suited Einstein just fine (Figure 8.2); it left plenty of time and energy to devote to working on aspects of physics that pleased *him*. Einstein could enjoy his domestic life, play his violin, socialize with the small group of loyal friends he had accumulated at the Poly, and theorize on the physics problems he thought important. He began by publishing several forgettable papers, but he soon got the hang of doing physics. Despite the birth of his son, Hans Albert, in mid-1904, and the resulting clamor in their small apartment during the first half of 1905, Einstein unleashed a flood of creativity almost unparalleled in intellectual history.

Those who become great scientists achieve their status because they solve the greatest problems, and Einstein's instinct for sniffing out and grappling with key problems was becoming unexcelled. During this creative period in 1905, Einstein began work on the most controversial and important problem of the day: proving that atoms exist and determining their weight. After sketching the

Figure 8.1 Einstein and Marić ca. 1903.

Figure 8.2 Einstein at the Patent Office, 1905. *Source:* Lucien Chavan/ETH Zürich.

development of atomic theory, we will describe this work and demonstrate what a *tour de force* his series of papers were in solidifying scientific opinion on the side of the existence of atoms and in accurately determining the weight of an atom.

It may surprise you to hear that even early in the twentieth century, some of the most sophisticated scientists doubted whether atoms and molecules actually existed or were merely a convenient fiction that allowed one to visualize the relative proportions of elements in different compounds. The idea of atoms had been speculative up to the time of Lavoisier. As we saw in Chapter 4, Lavoisier firmed up the idea of 33 elemental atoms incapable of division, with molecules being assembled from these atoms to form chemical compounds. Compounds differ from elements in that they, although sometimes with great difficulty, may be reduced once again to a mixture of elements.

In 1808, English chemist John Dalton fleshed out his insight that elemental and indestructible atoms should combine in fixed proportions to form chemical compounds [3]. Dalton set about the daunting task of inferring the relative weights of the atoms, which we know now as *atomic weights*, by careful evaluation of the relative weights of elements and compounds. We can illustrate how fraught with difficulty this process can be. Provisionally, since Dalton's measurements showed that hydrogen combined with about seven times its weight of oxygen, he anticipated that the oxygen atom was seven times the weight of hydrogen, assuming HO was the chemical formula for water [3, p. 138]. This initial assignment underscored the importance of getting both the relative weights and the *stoichiometry* right, because the correct formula is H_2O and the correct weight proportion is 16:1, not 14:1. Moreover, Dalton never considered the fact that hydrogen and oxygen, even though they were elements, formed only the diatomic elemental gases H_2 and O_2. Extending Dalton's concept to determine the relative weights of the known elements was the labor of many decades.

The work of the English pneumatic chemists led to another clue that helped to unravel the knotty problem of stoichiometry. In experiments with hydrogen and oxygen, Henry Cavendish noted that two volumes of hydrogen could combine (or rather, explode!) with exactly one volume of oxygen to make water [3, p. 109]. Other gases, among them NO and O_2, also combined in simple integral ratios. By 1810, Joseph Louis Gay-Lussac was able to formulate this principle: *The volumes of gas that appear or disappear in any reaction are in simple ratios to each other* [3, p. 163]. Note that this principle does not apply to the changes in *weights* occurring during chemical change because atomic weights can be large, nonintegral numbers.

In 1811, an otherwise obscure Italian scientist, Amedeo Avogadro, after meditating at length on Gay-Lussac's principle, concluded that its logical interpretation was this: *Equal volumes of different gases, under the same conditions of temperature and pressure, contain equal numbers of molecules* [3, p. 165]. His flash of brilliance probably developed like this: Molecules must be made up of atoms combining in simple ratios, and volumes of gas react in simple ratios, so we might legitimately expect that there are equal numbers of molecules in equal volumes of all gases. Even though this principle would prove to be true, this hypothesis, or perhaps the word *conjecture* is more appropriate, was far ahead of its time and did not garner rapid acceptance. However, it did extend the promise of determining the number of molecules in each unit volume of every gas, and hence the weight of every molecule. Little is known about the life of Avogadro except that he was a teacher of mathematics, but we trust that his surviving portrait did not do him justice (Figure 8.3).

Meanwhile, the traditional chemists who worked with solids and solutions had not been idle. By 1810, it was becoming accepted that hydrogen was the lightest element and that elements combined in fixed ratios to the weight of hydrogen, but the data on relative weights was scanty:

$$\text{hydrogen} = 1, \ \text{carbon} = 12, \ \text{oxygen} = 16, \ \text{chlorine} = 35.5, \ldots$$

Figure 8.3 Amedeo Avogadro, 1776–1856.

What were these chemists, groping for simplicity, to make of the relative weight of the chlorine atom being 35.5? And that is not experimental error or a typo—they got the relative weight right. It only became clear in the twentieth century that there are two abundant isotopes of chlorine; chlorine atoms always have 17 protons, but three-quarters of chlorine atoms have 18 neutrons and one-quarter have 20 neutrons. That averages out to a weight of 35.5 units. Neutrons weigh just slightly more than protons (about 0.1% more), whereas electrons weigh only 0.055% as much as protons.

These weight ratios listed above became known as *atomic weights.* Extending the concept of atomic weights to molecules is straightforward, and chemists could now be assured, for instance, that $35.5 + 1 = 36.5$ g of the gas HCl contained the same number of molecules as 12 g of carbon. These gram weights became known as *molecular weights*, or *moles* for short.

However, Nature had not made the problem of atomic weights simple for the pioneering chemists. For instance, it would have been helpful to have information about the atoms lying in weight between hydrogen and carbon, but they were all unknown at the turn of the nineteenth century. Element 2, helium, is a noble gas first discovered in the sun's atmosphere in 1868. Element 3, the very reactive metal lithium, was first separated in 1817. Element 4, beryllium, was first separated in 1828 but isolated in pure form only in 1898. Thus, many of these valuable pieces of the puzzle were missing.

Nevertheless, even missing these important pieces of the complex puzzle, chemists continued toiling at this essential but difficult conundrum. They had to sort through every family of chemical compounds and decide the stoichiometry of each. When this procedure was finished, chemists would have a progression of relative atomic weights starting from hydrogen at 1 unit and ranging to uranium at 238. In macroscopic terms, they knew that 1 g of hydrogen would contain the same number of atoms as 12 g of carbon or 55 g of iron. These atomic weights, according to Avogadro's hypothesis, are also the weights of a fixed volume of elemental monatomic gas. Therefore, 4 g of the monatomic gas He or 2 g of the diatomic gas hydrogen occupies a volume of about 24 liters at 1 atmosphere pressure and room temperature.

By 1869, enough atomic weights were in place for Dmitri Mendeleev to assemble an incomplete but still very useful periodic table, ordering the elements by atomic weights to group them into families having similar chemical properties [3, p. 349]. John Newlands and Lothar Meyer had presaged this, but Mendeleev actually predicted the discovery of germanium, gallium, and scandium, which corresponded to empty slots in his table. The fundamental classifying principle of atoms—atomic number, or the number of protons in an element's nucleus—was a concept that would only be clarified by Henry Moseley's X-ray studies in 1914, which were directly inspired by Niels Bohr's theory of the light emitted by excited hydrogen and helium atoms (Chapter 9).

So far, so good. Chemists had a grasp on the relative weights of most of the atoms on earth. The next step in proving atoms to be real, and not just a convenient abstraction, was to identify their true size. How many carbon atoms did that 12 g atomic weight of carbon contain? Avogadro's number, N, the number of molecules in a mole, is the parameter key to weighing the molecule because the weight of an individual atom m can be obtained from its atomic weight M by

$$m = \frac{M}{N}, \tag{8.1}$$

with the current value of Avogadro's number N being 6.022×10^{23} atoms per mole.

As for determining N, nineteenth century science did not have good options for examining the very small. Optical microscopes were highly developed, but it is hard to observe an atom of dimension near 0.1 nm (0.1×10^{-9} m) with visible light whose wavelength is between 300 and 500 nm. You might compare that to trying to handle a 2 mm pellet with forceps made from telephone poles.

But clues to the size of atoms were nevertheless observable. In 1827, botanist Robert Brown noted what must have been apparent to many others who used high-powered microscopes to examine particles below the size of 1 μm (1×10^{-6} m): They were in constant agitation. Brown noted that smaller particles ejected from pollen grains and even inorganic particles exhibited a random jitter. He even examined ground sphinx dust—perhaps looking for enhanced magical energy—and found the same jitter.

Physicists were beginning to appreciate in the nineteenth century that gas molecules themselves were in constant motion. German physicist Ludwig Boltzmann had developed a comprehensive theory that showed that the pressure exerted by a gas on the walls of its vessel was the result of bombardment of the gas molecules, and that each molecule of mass m in that gas had a kinetic energy ($1/2mv^2$, with v being the molecule's velocity) that was caused by and proportional to the temperature of the gas. (To make the proportionality work, you have to measure gas temperature on the absolute or Kelvin scale, as its zero point is not the freezing point of water, but $-273°$C.)

The earliest serious attempt at calculating the size of molecules seems to have been that of Loschmidt [4] in 1865. He used Boltzmann's concept that gases are comprised of very small colliding molecules and estimated the diameter of a gas molecule from a guess of what the density of liquefied gas should be. Surprisingly, his estimate was off by only one order of magnitude: It would give an Avogadro's number of 0.41×10^{23} per gram molecule.

Eight years later, James Clerk-Maxwell [5] noted that the velocities of gas molecules from Boltzmann's theory are of the order of hundreds of meters per second, whereas the diffusion coefficients of the same gases are below 1 cm^2/s. Maxwell vividly explained that if the molecules coming from a bottle of ammonia, which he opened with a flourish during a lecture, were not slowed down by an enormous number of random collisions, then people in the last row of the lecture hall would perceive their acrid odor instantly. Using Boltzmann's theory, Maxwell was able to estimate that Avogadro's number was 4.25×10^{23} per gram molecule, a big improvement over Loschmidt's estimate.

As Boltzmann's theory gained acceptance, speculation increased as to whether the intrinsic motion of gas or liquid molecules might be the origin of the Brownian motion. However, atoms had to be far, far smaller than the 1 μm dancing particles that Brown had observed. In fact, the diameter of a water molecule is near 0.1 nm (1 nm = 10^{-9} m).

In early 1905, the 26-year-old Einstein published the first of a series of six papers in which he derived (but did not necessarily evaluate) seven ways to determine Avogadro's number from four different physical phenomena. Although four of the methods are closely related to the Brownian motion, this *tour de force* demonstrated the reality of discrete atoms of a consistent size in such diverse ways that the atomic theory was credibly established to all but its most recalcitrant critics.

Einstein's first published paper [6] that included an evaluation of Avogadro's number, submitted in March 1905, was principally a calculation of the spectrum of cavity radiation and had nothing to do with the Brownian motion. At that time, the 5-year-old Planck radiation law (see Eq. (7.8)) was generally regarded as an only partially justified kluge that luckily managed to fit the entire spectrum of cavity radiation. The expression that we called the Rayleigh–Jeans radiation law giving the energy density in a cavity $\rho(\nu,T)$ at equilibrium at temperature T in a frequency interval near ν existed only in the skeletal form,

$$\rho(\nu, T) = C_0 T \nu^2, \tag{8.2}$$

with C_0 being a constant not yet theoretically derived. In that paper, Einstein actually was the first to derive this proportionality constant correctly, obtaining

$$\rho(\nu, T) = \frac{8\pi R T \nu^2}{c^3 N} \tag{8.3}$$

two months before Rayleigh published a calculation incorrect by a factor of 8 and three months before Jeans corrected Rayleigh's miscalculation [7]. In this equation, c is the velocity of light, T is the absolute temperature, N is Avogadro's number, and R is the gas constant.

If you skipped Chapter 7 on Planck, then you might wonder why the spectrum of cavity radiation should be related to Avogadro's number. Because the radiation is in equilibrium with the vibrating atoms of the wall, the energy of each resonant mode of radiation within the cavity is a multiple of the basic vibrational energy of an atom at the cavity wall: RT/N.

In one section of the paper, Einstein used this new expression at the long-wavelength limit in order to evaluate Avogadro's number. Although Planck in his 1900 paper had evaluated N by comparing experiment with the Planck radiation law, Einstein felt that his method was more firmly grounded theoretically and therefore publishable, so he evaluated the left-hand side of Eq. (8.3) from the experimental curve. On the right-hand side, the gas constant R and the speed of light c were known to good accuracy, allowing Einstein to solve for N. It is no surprise that

Einstein got the same value as Planck: 6.17×10^{23} molecules per mole. This was to be his most accurate determination when compared to the modern value of 6.02×10^{23} molecules per mole.

Using an improved modern value of the atomic weight for the H atom, 1.0076, Einstein's value for Avogadro's number would give the weight of the H atom as 1.63×10^{-22} g, compared to its accepted modern value of 1.67×10^{-22} g.

Einstein's next three papers showed that he had a sufficiently delicate touch to also extract the weight of a molecule from the microscopic jiggling observed by Brown. Einstein perceived that even though Brown's dancing particle may be far larger than the liquid molecules that constantly bombarded it, the Boltzmann theory is also applicable to it and requires that its energy is also proportional to the absolute temperature.

Einstein's extraction of the number of molecules in a mole from the Brownian motion is so clever that it almost seems magical [8]. Einstein started by noting that the n moles of Brownian particles, according to the "molecular-kinetic theory of heat," will bombard the walls of the container and be themselves bombarded with the faster-moving solute molecules. He showed that these particles should then exert an *osmotic* pressure P on the walls of their container that should satisfy the perfect gas equation

$$PV = nRT \tag{8.4}$$

if they are confined to a volume V at absolute temperature T. R is just the well-known gas constant. This law had also been recently shown by van't Hoff to also apply to dilute solutions or suspensions, such as the small particles observed in Brownian motion. Einstein also utilized key equations defining diffusion, which are set up in terms of density, ρ, expressed by the weight of particles dispersed in the solvent per volume,

$$\rho = \frac{Mn}{V}. \tag{8.5}$$

The number of moles n times the weight of each mole M gives the total weight of the particles in the volume V.

These small particles, battered on every side by molecules of water, do not inhabit a world described by Newton's second law of motion because particles in a liquid do not maintain a constant velocity when not being acted on by an exterior force. The resistive force slowing their motion through a fluid is related to the parameter *viscosity*, which can be measured by determining the resistance of a liquid to the motion of an object traveling through it. In the world of these particles, like the world of a swimmer who stops stroking or a paramecium that ceases the twirling of its flagella, bodies are rapidly slowed to a stop by the viscosity of the fluid. In a fluid of viscosity η, a sphere of radius a traveling at a velocity v encounters a resistive force F of magnitude

$$F = 6\pi\eta a v. \tag{8.6}$$

This relation was originally established by George Stokes in 1851. It is easy to verify by timing small metal spheres falling through a vertical tube of liquid.

Einstein imagined these Brownian particles to be pushed to the right by an unspecified force K on each particle as shown in Figure 8.4. This force might be the result of gravity or *osmotic pressure* engendered by the eagerness of the water molecules to dilute the dispersed Brownian particles. The force per unit volume can be obtained by multiplying this force by the density of particles and dividing by the mass of each particle from Eq. (8.1). (Following this argument will improve your ability to make unit changes!)

$$\frac{\text{force}}{\text{vol}} = \frac{\text{force}}{\text{particle}} \cdot \frac{\text{particles}}{\text{mole}} \cdot \frac{\text{mole}}{\text{mass of particles}} \cdot \frac{\text{mass of particles}}{\text{volume}} = \frac{KN\rho}{M}. \tag{8.7}$$

This force per unit volume establishes a pressure difference or gradient pushing the particles to the right to increase the pressure at the right side of the figure. Einstein carefully derives from thermodynamics an equation calculating that a volumetric force will cause a pressure change ΔP over every small distance Δx. This equation follows from the definition of pressure,

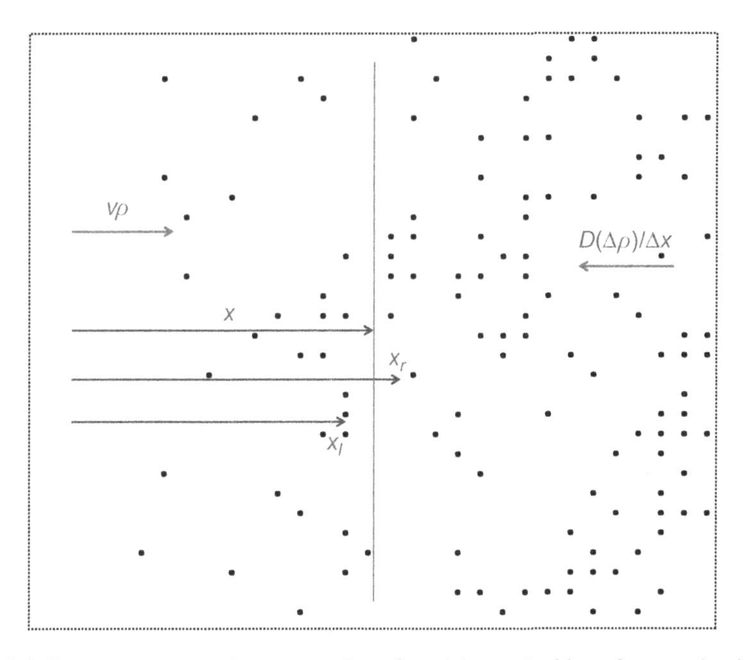

Figure 8.4 Computer-generated representation of particles pushed by a force to the right counterbalanced by a diffusive flow to the left.

$$\frac{\text{force}}{\text{vol}} = \frac{\Delta P}{\Delta x}. \tag{8.8}$$

Here we use the symbol Δ to mean the change from the right to the left side of a small interval (Figure 8.4),

$$\Delta P = P_r - P_l; \quad \Delta x = x_r - x_l. \tag{8.9}$$

Using the gas law (Eqs. (8.4) and (8.5)), we may eliminate pressure by expressing it in terms of density,

$$\frac{\Delta P}{\Delta x} = \frac{RT}{M}\frac{\Delta \rho}{\Delta x}, \tag{8.10}$$

again using the notation

$$\Delta \rho = \rho_r - \rho_l. \tag{8.11}$$

Since particles of radius a are pushed by a force K through a medium of viscosity η, they will reach a velocity v given by Stokes's law,

$$v = \frac{K}{6\pi\eta a}. \tag{8.12}$$

This velocity, multiplied by the mass of particles per unit volume ρ, gives the total flux (g/(s cm^2) of particles moved by K:

$$\text{Flux} = v\rho. \tag{8.13}$$

This flow to the right will compress the Brownian particles together; their increased density will establish a diffusive flow in the leftward direction that ultimately reaches the equilibrium flux

$$\text{Flux} = D\frac{(\Delta \rho)}{\Delta x}. \tag{8.14}$$

This equality is the defining equation of diffusion, named Fick's law. D is a constant called the *diffusion coefficient* that is characteristic of the particles and the solvent.

After this tedious description of how the Brownian particles move, Einstein shows how brilliantly these equations reduce into one very simplified expression. Substitute Eq. (8.12) into Eq. (8.13). Express the right side of Eq. (8.14) in terms of K by using Eqs. (8.7), (8.8), and (8.10) to give the equation

$$\frac{K\rho}{6\pi a\eta} = D\frac{M}{RT}\frac{K\rho N}{M}. \tag{8.15}$$

Eliminating the factors common to both sides yields

$$D = \frac{RT}{6a\pi\eta N}. \tag{8.16}$$

Einstein has reduced the two equations of fluid flow and diffusion to a single relation between the diffusion coefficient D of the Brownian particles and the number of molecules per mole, N. The temperature, liquid viscosity, and size of the particles are presumably known. The force K and the particle density ρ have canceled out. Now you see it, now you don't! The physical meaning of the disappearing K is that this force may have any cause: gravity, osmotic pressure, etc., and still be balanced by the countervailing diffusive flow.

Einstein now expanded the existing concept of diffusion in order to provide an experimental method of evaluating D. Figure 8.5 shows some computer-generated paths taken by a particle moving randomly in a plane. Einstein showed that the gross motion of each dancing particle was related to the physical problem of diffusion and that the diffusion coefficient could be expressed in terms of the root-mean-squared displacement of the particle, $[\langle X^2 \rangle]^{1/2}$, after the time t it has diffused from its starting point:

$$D = \frac{\langle X^2 \rangle}{2t} \tag{8.17}$$

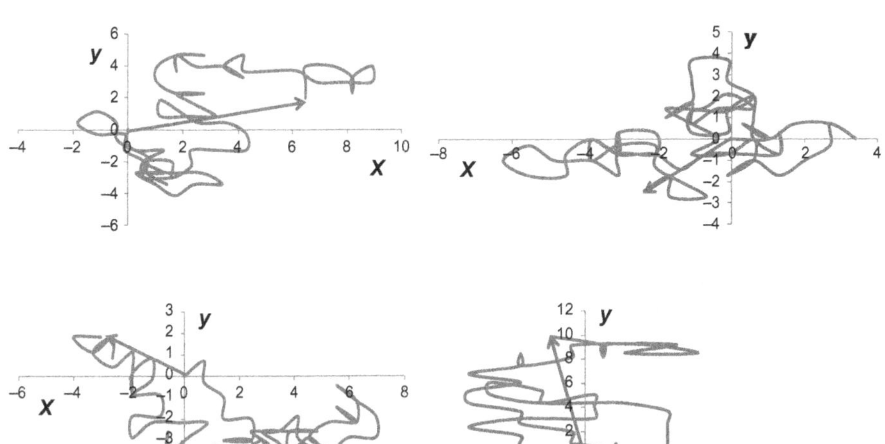

Figure 8.5 Four computer-generated diffusion paths. Each path is the result of 75 accumulated jumps in random directions beginning at the origin. The straight arrows directed away from the origin show the net motion. If the jumps were the result of linear motion, then the accumulated displacement would be 75 units instead of the 4–10 units displayed here.

The quadratic behavior in X is consistent with the idea that the particles may double back and not make the same progress they would at constant velocity when the particles would move at velocity v:

$$v = \frac{X}{t} \tag{8.18}$$

From experiments calculating the average root-mean-square particle-diffusion distance $[\langle X^2 \rangle]^{1/2}$ in time t, one can eliminate D, giving

$$N = RT \frac{t}{3\pi\eta a \langle X^2 \rangle} \tag{8.19}$$

Because Einstein knew every term on the right side could be determined, he could solve for Avogadro's number, N. This equation is really a stunning achievement. As Abraham Pais noted in his comprehensive and technical Einstein biography, *"Subtle Is the Lord,"* "one never ceases to experience surprise at this result, which seems, as it were, to come out of nowhere: prepare a set of small spheres which are nevertheless huge compared to simple (i.e., water) molecules, use a stopwatch and a microscope, and find Avogadro's number" [1, p. 97]. Well, perhaps not quite as simple as that sounds, because, as you can see from Figure 8.6, you have to average the diffusion distance of many random particle motions to get an accurate number for $\langle X^2 \rangle$.

Einstein had neither the facilities nor the inclination to do this experiment himself and concluded the paper with an invitation: "It is hoped that some enquirer may succeed shortly in solving the problem suggested here, which is so important in connection with the theory of heat." As if to entice experimenters to try these measurements, Einstein calculated that for particles of 1 μm radius suspended in water, the expected diffusion length after 1 min would be 6 μm.

More important and convincing than calculating the actual value for N was the fact that Einstein had clearly shown that the Brownian motion was a direct manifestation of the atomic nature of matter as interpreted by Boltzmann. Einstein himself was rather coy about relating his calculation to what Brown had observed because he had begun his groundbreaking paper with the disclaimer, "It is possible that the motions to be discussed here are identical with the so-called Brownian motion; however the data available to me are so imprecise that I could not form a judgment on the question." Perhaps this was simply youthful bravado trying to foster the notion that his theoretical work proceeded from deductive principles alone and was not primarily motivated by an attempt to clarify a mystery observed by experimenters.

Einstein's third determination of N was in his doctorial thesis, written in early 1905 but not published until 1906, titled "A New Determination of Molecular Dimensions" [9]. This paper examined the motion of a dilute solution of sugar in water. Einstein found experimental data for how sugar molecules increase the viscosity of the water in which they are dissolved, and from this he derived an equation relating the effective radius of the sugar molecules to N, Avogadro's number. You can see how these two quantities would be involved, since the product Na^3 is proportional to the volume of the sugar molecules in solution.

The diffusivity D of the big sugar molecules is the result of molecular collisions similar to those described previously for suspended particles in water and had been directly measured by others, so there was luckily no need this time to measure the random migration of the sugar molecules over a fixed period. The diffusion and viscous-resistance equations have the same form as in the preceding derivation and lead to Eq. (8.16), for which Einstein knew every parameter but N and a, the radius of the sugar molecules. With these two equations in two unknowns, he could evaluate N, giving 2.1×10^{23} molecules per mole. Luckily for Einstein, the editor of *Annalen der Physik* suggested new and better data for the viscosity of sugar solutions, raising Einstein's result to 4.15×10^{23}, and it was published with this improved value. Several years later, a French student pointed out an error in Einstein's calculation for how sugar increased the viscosity of water; the correction gave the improved result 6.56×10^{23}. This improved number was duly published and compares satisfactorily to the modern value of 6.02×10^{23}. As promised in his title, Einstein could also evaluate a, the radius of a sugar molecule, obtaining 62 nm, a result that he thought reasonable.

In these two methods of evaluating Brownian motion, Einstein had presumed the dancing particles to be in a thin horizontal layer and therefore little affected by gravity. Later in 1905, Einstein submitted a paper that specifically worked out the equations for how a gravitational field would concentrate particles at the bottom of a thin vertical volume [10]. He also worked out another independent case showing how the Brownian motion would rotate spherical particles. In this paper, he again estimated the appreciable effects that an experimenter might observe, in order to motivate evaluations of N by these two new methods. They have been carried out to good accuracy and are listed in Table 8.1.

Einstein noted in 1906 that the voltage noise between two plates of a charged capacitor is analogous to Brownian motion because it is also temperature dependent, but he did not do any evaluations himself.

Understandably, Einstein was too busy during this period of spectacular creativity to leave much of a historical record of his emotions. The closest he came was his oft-quoted letter to his friend Conrad Habicht. "So what are you up to, you frozen whale, you smoked, dried, canned piece of [sole]...? Why have you still not sent me your dissertation?" Einstein then described his work with zest and justifiable pride [12]:

> *I promise you four papers in return. The first deals with radiation and the energy properties of light and is very revolutionary,... The second paper is a determination of the true size of atoms.... The third proves that bodies of the order of magnitude 1/1000 mm, suspended in liquids, must already perform an observable random motion that is produced by thermal motion. Such movement of suspended bodies has actually been observed by physiologists who call it Brownian molecular motion. The fourth paper is only a rough draft at this point, and is an electrodynamics of moving bodies which employs a modification of the theory of space and time.*

Table 8.1 Determinations of Avogadro's Number

Method	Researcher	Year	Revised ($N/10^{23}$)	Original ($N/10^{23}$)
Molecular mean free path	Loschmidt	1865	0.41	0.41
Diffusion of gas molecules	Maxwell	1873	4.25	4.25
Cavity radiation spectrum	Planck	1900	6.17	6.17
Cavity radiation spectrum	Einstein	1905	6.17	6.17
Brownian motion of particles	Einstein	1905	6.4	None
Diffusion of sugar molecules in water	Einstein	1905	6.6	4.15
Vertical distribution in concentrated emulsions	Einstein	1905	6.8	None
Brownian rotation	Einstein	1905	6.5	None
Critical opalescence	Einstein	1910	7.5	None
Current accepted value		**2011**	**6.022**	

The column labeled "Original" gives the researcher's original published evaluation. The column labeled "Revised" gives the data available in 1923 and tabulated in Perrin's *Atoms* [11].

Even after these Brownian motion successes, Einstein was not willing to give up on this important problem until he had pounded on it one more time. In 1910, he turned to the work of a gifted Polish scientist named Marian von Smoluchowski [13], who had explained why liquids become "opalescent" or milky as they are heated through their boiling point. Smoluchowski had begun with a statistical analysis that demonstrated that the smaller the volume of fluid examined, the more likely it is to have a larger or smaller density than its bulk value. As molecules zig and zag into a very small volume, they temporarily make its density higher or lower, and the smaller the volume you are examining, the more likely it is to depart from density equilibrium.

That is not a very profound idea so far, but Smoluchowski demonstrated that in the vicinity of the boiling point the out-of-equilibrium regions should become much bigger. At most temperatures, it is characteristic of liquids that departures from the equilibrium density are quickly damped out, but measurement had shown that liquids became extremely compressible near their boiling point. Smoluchowski was able to calculate that the out-of-equilibrium regions near the boiling point of water should therefore grow to approach the ½ μm wavelength of light, so they should scatter visible light. Actually, this phenomenon had been observed and christened *critical opalescence*.

Einstein [14] calculated the fraction of light scattered at right angles to a light beam as a function of water's refractive index (a measure of how much water bends light rays) and compressibility near the boiling point. Just as in the diffusivity equations given previously, the factor RT/N is a parameter because it is proportional to the square of each molecule's velocity and molecular motion causes critical opalescence. Once again, Einstein was able to solve for N, with data obtained from a later experiment giving an N of 7.5×10^{23} molecules per mole.

Incidentally, these concepts are closely related to the calculation published by Lord Rayleigh [15] in 1871 that attributed the blue of the sky to scattering of the shorter (blue) wavelengths of sunlight by density fluctuations in the upper atmosphere. Density fluctuations are larger in size in the low-pressure upper atmosphere, and the shorter blue waves are more likely to be scattered by small regions of anomalous density. Such scattering is improbable, but the volume of sky is large and the scattered blue light accumulates to give a clear sky a decidedly bluish tint.

By contrast, the sunset has a reddish tint because much of the blue light has been scattered as sunlight skims through the atmosphere before reaching our eyes. We frequently can observe that atmospheric particles or water droplets can enhance the reddish color of the sunset when they are in abundance because large particles preferentially scatter the longer wavelength red and orange rays from the sun to our eyes.

When Einstein had completed this work, he had produced overwhelming evidence for the concrete existence of atoms and encouraged a consistent series of determinations of Avogadro's number. Table 8.1 summarizes these results. The column listing the "Revised $N/10^{23}$" values is convincing that these measurements converge to a value close to what more accurate methods have given us today. The column labeled "Original $N/10^{23}$" reminds us that at the time of publication, the data were less compelling; in some cases the data had not been acquired yet, whereas in others it took several years for a consistent experiment to be completed.

Jean Perrin [11] neatly summarized these determinations in his landmark book *Atoms*, along with other methods based on radioactive decay which we have not discussed.

As Einstein finished these calculations, which laid a firm foundation for the atomic theory, his imagination and depth of perception flourished enough to clarify problems that as yet were hardly recognized. We will just skim over the surface of three of them for the purpose of completing this sketch of Einstein's life.

Immediately after publishing his work on the Brownian motion, Einstein jolted the scientific community with his theory of special relativity, another work published in the miraculous year 1905. This theory addresses once again the problem of the "aether," the hypothetical substance filling all space that electromagnetic waves can set into oscillation in order to propagate. Physicists at the end of the nineteenth century speculated that this aether should define a universal state of rest in the universe and earth's motion through the aether should be detectible with optical instruments. However, in 1887, Michelson and Morley measured to great accuracy that the speed of light traveling in the direction of earth's orbital motion was equal to the speed of light traveling perpendicular to earth's motion. Michelson, incidentally, was totally perplexed by the results of his experiment, which he had designed to deduce the speed of earth relative to the aether. Einstein later gave conflicting accounts over whether he had been familiar with this work or had derived relativity theory from conceptual grounds alone [1, p. 172].

However, it was typical of Einstein's intuition that he accepted the accumulating evidence that there was no reference frame "at rest" to which all motion could be observed, and he deduced deep and useful consequences from the two very simple postulates:

1. The equations of physics are the same in all unaccelerated reference frames.
2. The speed of light is the same in all reference frames, no matter how rapid their relative motions, as long as they are not accelerating.

The first seems rather like the way the universe should be constructed, but the second, the constancy of the velocity of light, is simply not intuitive. That the light from our sun as we travel toward a nearby star moves with the same velocity as the light that is propagated back to a star from which we are receding seems baffling. This would not be the behavior, for example, of a stone thrown from a moving boat: The forward velocity of the stone would be augmented by the boat's speed, whereas the velocity of a stone thrown backward would be diminished by the boat's speed.

In 1632, Galileo [16] had considered this sort of relativity in his *Dialogues Concerning the Two Chief World Systems*. Arguing in favor of a rotating spherical earth, Galileo noted that an observer below deck on a smoothly sailing ship cannot determine whether or not the ship is in motion on the surface of the water. If he hops in the air, he will return to the same spot on the cabin floor. Butterflies will flutter about the cabin or fish will swim in a bowl resting on the cabin floor independently of the ship's relative motion.

To formulate his view of this problem, Einstein [17] originated the theory of relativity to describe the behavior of events observed in coordinate systems moving with respect to each other. Imagine that one coordinate system, x, y, and z, is at rest while the second moves to the right at constant velocity v parallel to the x coordinate (Figure 8.6). These axes can be labeled x', y', and z'. Events take place at a certain time t or t' in the primed or unprimed coordinate systems.

Consider how the instantaneous flash from a firecracker exploding at $t = 0$ would propagate away from the origin in the fixed reference frame. The flash would be described by a spherical locus of points expanding with time at the velocity of light c:

$$x^2 + y^2 + z^2 = c^2t^2. \tag{8.20}$$

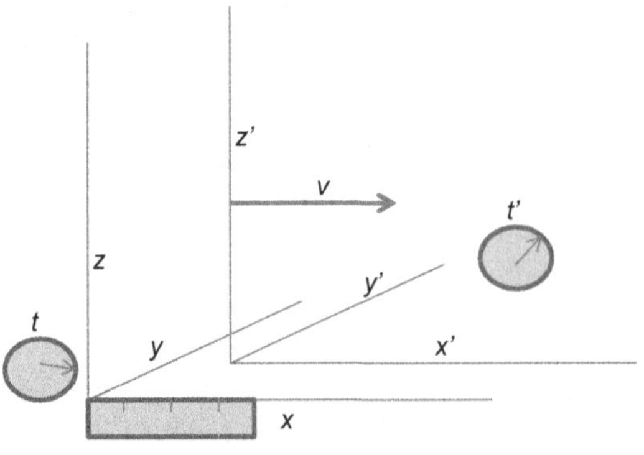

Figure 8.6 Two reference frames. The primed frame moves to the right with constant velocity v with respect to the unprimed.

Suppose a second reference frame moves to the right (Figure 8.6) at velocity v in the x direction but synchronized exactly so that the two origins coincide at the instant of the explosion.

According to Einstein's two postulates listed above, an observer moving in the primed reference frame would also see a spherical light wave expanding from the origin in the primed frame so that the locus of points defining the flash would satisfy

$$(x')^2 + (y')^2 + (z')^2 = c^2(t')^2. \tag{8.21}$$

It is becoming clear that both these two simultaneous equations can only be true if something very weird happens to the local space and time in these two reference frames. A reasonable simultaneous solution would take into account the velocity in the x direction and the time difference due to the relative motion and might be of the form

$$x' = \gamma(x - vt), \quad y' = y, \quad z' = z, \quad t' = \gamma\left(t - \frac{xv}{c^2}\right). \tag{8.22}$$

By substituting this equation into Eqs. (8.20) and (8.21), Einstein was able to solve for γ to derive a transformation that satisfied all of these conditions:

$$x' = \frac{x - vt}{\sqrt{1 - (v^2/c^2)}} \tag{8.23}$$

$$y = y', \quad z = z' \tag{8.24}$$

$$t' = \frac{t - vx/c^2}{\sqrt{1 - (v^2/c^2)}}. \tag{8.25}$$

This transformation, although apparently unknown to Einstein, had been previously published by Hendrik Lorentz [18], who had developed it over many years by considering how Maxwell's equations of electrodynamics transform between reference frames in relative motion.

Three surprising consequences leap out of these equations, showing us that relativistic space and time are disturbingly different from Galilean space and time.

From Eq. (8.25): Events that occur at $t = 0$ in the unprimed coordinate system do not necessarily occur at $t' = 0$ in the primed coordinate system. Only events at $x = 0$ will be simultaneous in both frames. This is called the *relativity of simultaneity*.

Also from Eq. (8.25): A clock at rest in the unprimed coordinate system at $x = 0$ will tick at intervals

$$t' = \frac{t}{\sqrt{1 - (v^2/c^2)}} \tag{8.26}$$

in the primed coordinate system so that a second will have a longer duration than in the coordinate system at rest. This effect is called *relativistic time dilation*. It is an encouraging result for our dreams of exploring the universe because it allows a traveler in a rapidly moving rocket ship to explore much further in his lifetime than one might expect. Another consequence is that, with respect to earth, one would expect to age less on a rapidly moving rocket ship than in the unprimed reference frame one left behind. This leads to the *twins' paradox* in which one twin who travels at an appreciable fraction of the speed of light and then returns to earth will have aged less than his stay-at-home sibling.

Remember, there is no fundamental difference between the moving and fixed reference systems. The entire derivation would be unaltered with the primed system at rest and the unprimed system moving to the left at velocity v. It is all relative! The twin in the rocket would not feel that his seconds were ticking by more slowly than the seconds back on earth. Does the confrontation of the twins of different ages back on earth after the rocket ride falsify this assertion? No, because the massive accelerations undergone by the rocket-riding twin are outside the assumptions of the special theory of relativity; the later general theory of relativity takes the acceleration into account and properly confirms a true age difference.

Another consequence of relativistic kinematics is that measuring rods at rest in the unprimed reference frame appear to shrink when viewed from the primed system. Imagine a rod of length x in the unprimed coordinate system. Any measurement of the length of the rod in the moving primed coordinate system must be made at the same time t' in this coordinate system. The measurement of the length of a rod must therefore be made in the primed system, from whose point of view

$$x = \frac{x' + vt'}{\sqrt{1 - (v^2/c^2)}} \tag{8.27}$$

The length of the rod measured at time t' in the primed system will thus be

$$x' = x\sqrt{1 - \frac{v^2}{c^2}} \tag{8.28}$$

and the length of the rod measured in the moving system will be smaller than it would be at rest.

Another consequence Einstein extracted from these transformations was the famous

$$E = mc^2 \tag{8.29}$$

which he derived by using the Lorentz transformation to calculate how the energy of a mass would change after emitting light waves.

As you can see, to apply the special theory of relativity correctly, you must pay very close attention to what happens in reference frames in relative motion. This theory solved the problem of how light can propagate at constant velocity in all reference frames at the expense of changing our most cherished intuitive notions about the independence of time and space. The twins' paradox alone might have killed a less-compelling theory. Special relativity was originally met with mixed reviews but has since been supported by many observations on the lifetimes of subatomic particles, the behavior of electrons in atoms, and the conservation of matter and energy.

In the final section of the 1905 paper in which Einstein [6] derived his formula for the emission of cavity radiation at long wavelengths, he proposed a solution to a completely different physics mystery: the photoelectric effect. The photoelectric effect occurs when light falls on a metal surface and ejects electrons. The apparatus to study this effect somewhat resembles a Crookes tube: It constitutes a metallic electrode that can be bombarded by light rays of varying frequencies (colors) and intensities and another electrode that could be set to a more positive voltage to collect the ejected electrons (Figure 8.7). With such apparatus, one can even determine the energy of the ejected electron by setting the potential between the electrodes to repel the slower and less energetic electrons. The experiment must be done in high vacuum, ideally on an atomically clean metallic surface.

Classically, that is according to Maxwell's equations, one would expect that when the intensity of the light was increased, more electrons would be ejected. After all, if the light were falling on your hand, then more-intense light would warm your hand more. But that is not what happened at all. In 1902, Phillip Leonard [19] arranged a Crookes tube (with a simpler anode shape than Figure 8.7) so that it could be bombarded by light from a carbon arc whose intensity and frequency could both be controlled. To his surprise, he found that low-frequency reddish light would not eject electrons no matter how much light energy bombarded the surface. At the other extreme, high-frequency violet light ejected some electrons even at very low light intensity. In fact, the energy of the ejected electrons increased linearly with light frequency.

To Einstein, this relationship between frequency and energy was reminiscent of Planck's theory of cavity radiation where he had found that inside a hot cavity the

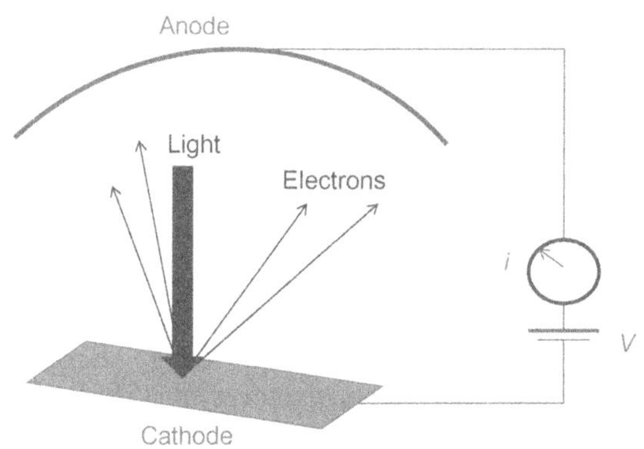

Figure 8.7 The light beam ejects photoelectrons that can be collected by the positive anode and detected at the galvanometer. The kinetic energy of electrons can be determined by making the voltage increasingly negative so that the slower electrons cannot strike the anode.

electromagnetic radiation possessed energy proportional to it's oscillation frequency. This whole riddle became perfectly understandable if the energy contained in each light wave was proportional to its frequency, ν. However, one had to add a further radical hypothesis to explain Leonard's result, that each light wave transferred its entire energy to one and only one electron.

Einstein's expression for the energy of a photoelectron had to account for the energy barrier ø that the photoelectron must surmount to escape from the metal; ø was thus a binding energy of electrons characteristic of each metal, now called the *work function*, which reduced the ejected electron's kinetic energy to

$$E = h\nu - \phi. \tag{8.30}$$

This equation has the required form to explain Leonard's measurements. It even offered a new way of measuring Planck's constant h: from the slope of the electron energy versus frequency curve. Nevertheless, this radical theory took awhile to be accepted. The conservative Planck resisted extending his expression

$$E = h\nu \tag{8.31}$$

to light because excellent evidence existed that light was a wave phenomenon and not a collection of particles. Understanding the wave–particle duality of light took many years to sort out, but Einstein was eventually awarded the 1921 Nobel Prize for this work on the photoelectric effect.

These achievements of 1905 soon lifted Einstein out of patent office obscurity. By 1908, he became a lecturer at the University of Bern, and academic life agreed with him so well that he resigned from the patent office to become a docent at the University of Zurich the following year. In 1911, he accepted a full professorship at Karl-Ferdinand University in Prague.

During these years, Einstein had a deep and incisive illumination about the nature of gravitation. The accepted theory, by Newton, was that every piece of matter of mass m in the universe attracts every other piece M with a force proportional to the product of their masses and inversely proportional to the square of their separation, r,

$$F = G\frac{mM}{r^2}, \tag{8.32}$$

where the universal constant G is a very small number representing the force attracting two unit masses separated by one unit distance. This theory predicted the planetary orbits remarkably well, with some minor problems, such as the advance of Mercury's perihelion, which we will discuss shortly. What seemed mysterious about Newton's theory was that it implied (Chapter 5) that every bit of matter felt the presence of every other bit, even those located immense distances away. Newton could not supply a rationale for this *action-at-a-distance*, and the scientific world learned to live with this slightly mysterious aspect of a theory that was, even considering this flaw, the most powerful the mind of man had ever devised.

Einstein's deep perception about gravitation was that being attracted by gravity is completely equivalent to being accelerated. His example, or *thought experiment*, was very illuminating: a person in an elevator cannot distinguish whether the elevator is being uniformly accelerated by a force on the cable or is at rest in a gravitational field. Devising a consistent theory of gravitation from this insight took many years of searching for the correct mathematical form. Einstein did not consider himself to be a particularly gifted mathematician, and for this work he had to assimilate and use some fancy geometry to show how gravitation is equivalent to warping space in the vicinity of each mass. One might visualize (two-dimensional) space as a taut rubber membrane. Masses set on it would cause it to sag. Heavier masses force it to sag deeply enough so that they strongly "attract" nearby masses to slide down the incline.

Einstein's new general theory of relativity explained the puzzling advance of the perihelion of the orbit of Mercury. Newton's gravitational theory predicts that a single planet orbiting a dense sun should traverse an elliptical orbit whose perihelion, its closest approach to the sun, should remain at a fixed point in its orbit. Corrections have to be made for the influences of other planets and tidal forces from the sun, but it had been observed in the mid-nineteenth century that Mercury's perihelion was advancing by an inexplicably large amount beyond what these corrections could explain. Possible explanations might have been an unobserved planet or asteroid swarm inside Mercury's orbit or Venus being significantly heavier than calculated. None of these explanations proved viable. In 1916, Einstein, using his newly fashioned general theory of relativity, which incorporated the distortion in space caused by our massive sun, correctly calculated the 43″/century unaccounted-for advance. Newtonian mechanics and gravitation were just not able to explain this orbital anomaly, but Einstein's general theory of relativity was. This success was central to the rapid acceptance of general relativity.

After deriving this result and solving a vexing 50-year-old puzzle, Einstein felt "beside myself with joyous excitement" and thought he had felt "palpitations of the heart" [1, p. 253].

For most of Einstein's lifetime, this general theory of relativity was regarded as an exotic creature having little application to most problems and not susceptible to many tests. One further meaningful test, however, might be done. Newton's theory of gravitation, as exemplified by Eq. (8.32), does not predict that massless light beams would be deviated by the gravity of even the heaviest of objects. However, Einstein's theory predicts that light beams would bend as they followed the curvature imposed on space itself by a nearby heavy mass [20]. Einstein calculated that this effect should be detectable for starlight passing near to the sun (Figure 8.8), but the measurement of a slight deviation in the apparent position of a star as its light skimmed the solar disk required observation during a total solar eclipse. The measurements were made by Sir Arthur Eddington on an island off the West African Coast in 1919 and deemed to confirm Einstein's theory, making headlines around the globe. Current reevaluations of Eddington's measurements show that they should have been considered to be ambiguous rather than definitive. The enthusiasm for Einstein's general relativity, however, carried the day; his theory

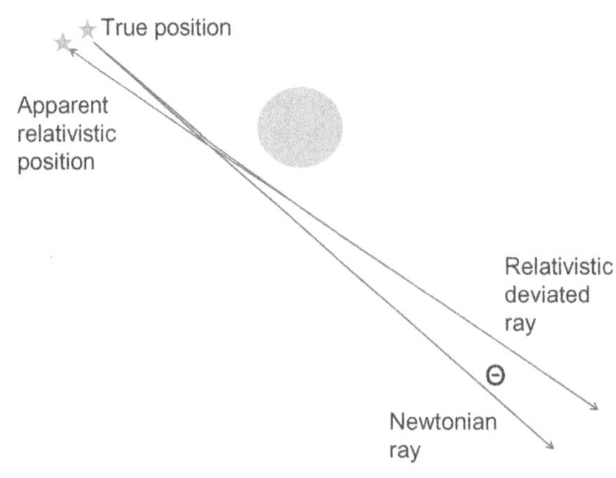

Figure 8.8 Test of the general theory of relativity. Far away from the sun's disk, the light from a star will follow the straight path marked "Newtonian ray." As light from the star approaches the sun's disk, it will follow the "deviated ray" trajectory as it is bent toward the sun by gravitation. The apparent star position will accordingly move by the angle Θ.

just "smelled" better than Newton's. Since that time, however, general relativity has been put to many stringent tests, including the functioning of our current global positioning system, without falsification.

Meanwhile, Einstein had been climbing the academic ladder in Germany. In 1914, he accepted a prestigious joint appointment as director of the Kaiser Wilhelm Institute in Berlin and a professorship at Berlin's Humboldt University. This gave Einstein the opportunity to terminate the increasingly unhappy marriage to Marić, who remained in Zurich with Hans Albert and a second son, Eduard. Einstein had been developing a relationship with his cousin Elsa, whom he married in 1919.

Einstein lived in Berlin during the painful years of the First World War and the turbulent years after the armistice of 1918, but by 1932 it was clear that the new Hitler regime intended to ratchet anti-Semitism to dangerous levels. Einstein who was visiting the United States at the time, decided to remain in America.

Einstein went on to devote a long and productive life to science (Figure 8.9). He extended the implications of general relativity to cosmology. He was a significant and perceptive critic of the developing quantum theories. His refusal to except the indeterminacy built into wave mechanics somewhat marginalized him, but the arguments he advanced definitely helped to sharpen the focus of the emerging quantum theory.

Einstein's great prestige occasionally propelled him into politics. He agreed to sign a letter to President Roosevelt calling his attention to the possibility of constructing an atomic bomb and leading to the Manhattan Project. Perhaps because of his many friendships with Jewish scientists escaping the holocaust, Einstein's interest in Jewish causes and the newly emerging state of Israel became significant. He had always been a pacifist, but the heightened dangers of a nuclear armed world coupled with regret over his own indirect role in originating the atomic bomb made him even more devoted to the cause of peace.

His final years at Princeton's Institute for Advanced Studies were spent on the still illusive *unified field theory* that sought to generalize his theory of gravitation by

Figure 8.9 Einstein lectures in Vienna in 1921. *Source:* F. Schmutzer.

incorporating the laws of electodynamics. Later, this work matured into a world wide and still unfinished effort to generalize the four forces known to physicists: gravity, electromagnetism, the weak nuclear force that underlies β-decay, and the strong force that holds nuclei together despite electrostatic repulsion between the protons.

Einstein died in 1955 after refusing surgery for an aortic aneurism. The scientist who sought and prized elegance in his theories did not want to depart this world in ugly dependence. "I have done my share, it is time to go. I will do it elegantly." He left behind a matchless body of scientific work that still astonishes us with its fertility of imagination and depth of conception.

References

[1] A. Pais, "Subtle Is the Lord...": The Science and the Life of Albert Einstein, Clarendon Press, Oxford, 1982, p. 40.
[2] W. Isaacson, Einstein: His Life and Universe, Simon and Schuster, New York, NY, 2007, p. 47.
[3] W.H. Brock, The Norton History of Chemistry, Norton, New York, NY, 1993, p. 128
[4] J. Loschmidt, Zür Grösse der Luftmolecüle, Sitzungsberichte der Kaiserliche Akademie der Wissenschaft, Wien 52 (2) (1865) 395–413.
[5] J.C. Maxwell, Molecules, Nature 8 (204) (1873) 437–441.
[6] A. Einstein, On a heuristic point of view considering the production and transformation of light, Annalen der Physik 17 (1905) 132. An English translation is available, in J. Statchel (Ed.), Einstein's Miraculous Year, Princeton University Press, Princeton, NJ, 1998, pp. 177–198.
[7] J.W.S. Rayleigh, The dynamical theory of gases and radiation. Nature 72 (1905) 54–55. J.H. Jeans, On the partition of energy between matter and aether, Phil. Mag. 10 (1905) 91–98.

[8] A. Einstein, On the motion of small particles suspended in liquids at rest required by the molecular-kinetic theory of heat, Annalen der Physik 17 (1905) 549 An English translation is available, in: J. Statchel (Ed.), Einstein's Miraculous Year, Princeton University Press, Princeton, NJ, 1998, pp. 85–98.

[9] A. Einstein, A new determination of molecular dimensions, Annalen der Physik 19 (1906) 289. An English translation is available, in: J. Statchel (Ed.), Einstein's Miraculous Year, Princeton University Press, Princeton, NJ, 1998, pp. 45–65.

[10] A. Einstein, On the Theory of Brownian Motion. Annalen der Physik 19 (1906) 371–381.

[11] J. Perrin, Atoms (D.L. Hammick, Trans.) second English ed. revised, Constable, London, 1923.

[12] W. Isaacson, Einstein: His Life and Universe, Simon and Schuster, New York, NY, 2007. p. 93.

[13] M. von Smoluchowski, Molecular-Kinetic Theory of the Opalescence of Gases at the Critical Point. Annalen der Physik 25 (1908) 205–226.

[14] A. Einstein, The Theory of Opalescence of Homogeneous Liquids and Mixtures near the Critical Point. Annalen der Physik 33 (1910) 1275–1298.

[15] Lord Rayleigh, On the light from the sky, its polarization and colour. Phil. Mag. 41 (1871) 107–110.

[16] G. Galilei, (S. Drake, Trans.) Dialogue Concerning the Two Chief World Systems, University of California Press, Berkeley, CA, 1953.

[17] A. Einstein, Annalen der Physik 17 (1905) 895 An English translation of "On the Electrodynamics of Moving Bodies" is available in Einstein's Miraculous Year, in: J. Statchel (Ed.), Princeton University Press, Princeton, NJ, 1998, pp. 123–159.

[18] H.A. Lorentz, Electromagnetic phenomena in a system moving with any velocity smaller than that of light. Proc. K. Ak. Amsterdam, 1904, 6, pp. 809–831.

[19] P. Lenard, M. Wolf, On the photoelectric effect, Annalen der Physik 8 (1902) 149–198.

[20] H.A. Lorentz, A. Einstein, H. Minkowski, H. Weyl, (W. Perrett, G.B. Jeffery, Trans.) The Principles of Relativity, Dover, New York, NY, 1913.

Bibliography

Annalen der Physik
The 1905 papers are available on the Internet in English and German.

W.H. Brock, The Norton History of Chemistry, Norton, New York, NY, 1993.
Technical and excellent; offers an exhaustive description of the development of atomic weight.

W. Isaacson, Einstein: His Life and Universe, Simon and Schuster, New York, NY, 2007.
Very readable, but skimpy on Brownian motion and less technical.

A. Pais, "Subtle Is the Lord...": The Science and the Life of Albert Einstein, Clarendon Press, Oxford, 1982.
Complete with all the source material references, technical details, and equations.

J Perrin, (D.L. Hammick, Trans.) Atoms, second English ed. revised, Constable, London, 1923
Hard to find, but a gem.

J. Statchel (Ed.), Einstein's Miraculous Year, Princeton University Press, Princeton, NJ, 1998.
English translations of five of the 1905 papers with helpful perspectives.

9 Niels Bohr Models the Hydrogen Atom as a Quantized System with Compelling Exactness, but His Later Career Proves that Collaboration and Developing New Talent Can Become More Significant than the Groundbreaking Research of Any Individual

Sometimes, great scientific progress is forged in solitude by lone geniuses. Newton's development of calculus and his theory of gravitation, for example, seem to have been completed without a smidgeon of collaboration and set aside for years without publication or sharing with a single soul. Another example we have described in this book is Röntgen's discovery of X-rays, which he did not share with even his closest colleagues until his work was completed and published.

But there are times when new concepts arise that are so complex and multifaceted that progress is best made with many judicious inputs, incorporating data from a multitude of workers through dialogue and questioning. Although Niels Bohr's career began with a brilliant independent discovery, his greatest contribution to twentieth-century physics came when he served as a sort of discussion guide and umpire, nursing into bloom the tender shoots of quantum mechanics. Bohr achieved his status as the referee of quantum mechanics because it was clear to all of the major participants that his love of truth transcended his personal interests. His modesty made it easy for him to give credit to others, so he developed trusted collaborators and students around the world.

Niels Bohr was born in 1885 in a mansion across the square from Copenhagen's Christianborg castle to a happy, affluent, and successful family. Both his father and his mother were from accomplished lineages that included scientists, engineers, teachers, judges, and theologians. His father, Christian, was a professor of physiology who had made useful discoveries on oxygen transport in blood. Niels grew tall and strong, tussling frequently with his younger brother, Harald, and ultimately

How the Great Scientists Reasoned. DOI: http://dx.doi.org/10.1016/B978-0-12-398498-2.00009-6

becoming a serious soccer player. He acquired a reputation for being somewhat distracted and clumsy. In school, he was generally attentive and had a gift for memorizing lyric poetry, but as he approached the end of his secondary school career, he chose to specialize in mathematics where he soon developed formidable mastery.

When he entered Copenhagen University, young Niels chose physics as his major, doubtless with guidance from his father. He was successful enough in his studies to win a prize for a paper describing an exotic method for determining the surface tension of liquids. For his master's degree topic, he was challenged to derive the properties of metals by considering their electrons to be a classical gas, and he expanded these ideas for his PhD thesis. Unfortunately, this topic really just explored the sterility of the classical (non-quantum mechanical) approach and was of only marginal interest to his fellow physicists [1]. Moreover, he was delayed in getting his degree by a problem that plagued him all his life: an obsessive need to endlessly revise his manuscripts before submission. For most of us, 14 revisions would seem too much, but Bohr thought it necessary to sweat bullets to express his ideas as clearly as possible. His photograph as a young man shows a composed face and a gaze focused on a far horizon (Figure 9.1).

Bohr's adviser, Professor Christianson, presided at Bohr's thesis defense in 1911, lamenting that the thesis was not available in a foreign language and remarking that no one in Denmark was really qualified to criticize it. This must have emphasized for the ambitious young man that he would have to go abroad to gain perspective on the significant problems opening up in physics.

The newly minted Dr. Bohr now reached for the top: He set off for Cambridge's Cavendish laboratory and studied with J.J. Thomson (Nobel prize in 1906), the discoverer of the electron. This was a time of great ferment and rapid development in atomic physics, and Thomson had been a leader in identifying the electron and arguing that its mass was less than 1/1000th that of the positive portion of the atom [1, pp. 117–120]. Thomson could be quite imaginative; two decades earlier he had published a prize-winning but gloriously incorrect paper maintaining that atoms resembled fluid vortices. In 1911, Thomson's favorite picture of the atom was his *plum-pudding model* that distributed the positive charge as a uniform "pudding," with the electrons circulating through it as "plums."

Figure 9.1 Niels Bohr as a young man.

As soon as he arrived in Cambridge, Bohr discovered that Thompson's personal style, formal, aloof, and rather distant, was a poor match for him. The naïve Bohr began their relationship by trying to explain in his labored and fractured English why some of Thompson's calculations were wrong, anticipating that Thompson would receive the corrections warmly and join him in doggedly straightening out the problems. Thomson was not about to be corrected by an untried young foreigner, and he avoided Bohr thereafter. Bohr made the best of the Cambridge environment for a few months and then arranged to transfer to the laboratory of Earnest Rutherford.

In moving to Rutherford's laboratory in Manchester in early 1912, Bohr demonstrated his nascent talent for identifying revolutionary research and productive collaborators. Rutherford was a vigorous, bluff, and loud New Zealand bull devoid of stuffiness—a man you could talk to and dispute with. Rutherford had first identified the three "rays" emitted by naturally radioactive elements: α-, β-, and γ, and showed that α-rays were energetic helium nuclei. Along with his collaborator, the brilliant chemist Fredrick Soddy, he had done laborious but groundbreaking work to identify the origins of natural radioactivity by sketching out the chain of transmuted elements that radium, thorium, uranium, and actinium decay into before becoming stable atoms.

Rutherford must have had more than a little of the arrogant physicist in him, though (Figure 9.2), because he inspired his research team at Manchester with with the exhortation "all science is either physics or stamp collecting." It was a well-deserved irony that when he won the Nobel Prize in 1908, it was in chemistry for his work on radioactive transmutation.

But the greatest of Rutherford's achievements was to invent a technique for how atomic particles were to be probed and investigated for the next century: by scattering other energetic particles from them. At that time, scientists had no effective way of looking inside the atom to test their hypotheses about its structure. Rutherford's inspiration was to use heavy and energetic α-particles to determine the size of the nucleus. A ductile metal such as gold could be beaten into foil only a few atoms thick. By directing a beam of α-particles through the gold and counting the scintillations they produced as they impacted different positions on a zinc sulfide screen, he could determine what fraction of α-particles were scattered through large angles (Figure 9.3). Rutherford and his grad student Hans Geiger

Figure 9.2 Lord Rutherford.
Source: Library of Congress.

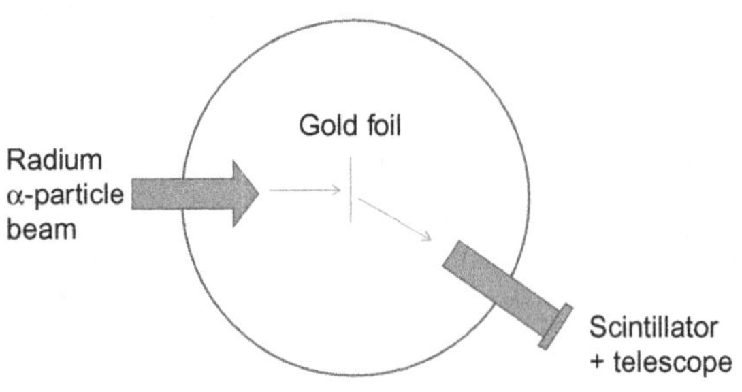

Figure 9.3 Geiger–Marsden apparatus. Alpha particles from radium are scattered from the gold foil and detected by the scintillator screen. The telescope slides around the circular track; the radium source is fixed.

were expecting very little scattering of the dense and energetic α-particles by the smeared out charges in plum-pudding atoms.

Consider an example of what one might learn from a simpler scattering experiment. Suppose it is a pitch black night and you are told that you are standing a few meters away from a thin screen of vegetation of an unknown type. Using as your only tools an ample supply of tennis balls, you are asked to determine what type of vegetation lies in front of you. You peg a few balls into the night and just hear a swishing sound as they travel through the vegetation almost without deflection. Perhaps the vegetation is just grasses or thin bushes, you begin to think. After throwing about 100 balls, however, you hear a solid "thunk" and a ball bounces back directly at you. You would justifiably suspect you had hit the trunk of a tree and would be able to infer that there were at least a few trees out there.

Similarly, Geiger and his undergraduate assistant Ernest Marsden were stunned to observe that a miniscule fraction of the α-particles—fewer than 1 in 10,000—bounced directly backward after colliding with the gold foil [2, p. 317].

When Rutherford saw the results, he enthused [2]

> *It was quite the most incredible event that has ever happened to me in my life. It was almost as incredible as if you fired a 15-inch shell at a piece of tissue paper and it came back and hit you. On consideration, I realized that this scattering backward must be the result of a single collision, and when I made calculations I saw that it was impossible to get anything of that order of magnitude unless you took a system in which the greater part of the mass of the atom was concentrated in a minute nucleus. It was then that I had the idea of an atom with a minute massive center, carrying a charge.*

The unexpectedly small nuclear diameter Rutherford calculated for gold from electric theory was near 10 fm (10×10^{-15} m), about 1/100,000th the diameter of an atom. If the gold atom were expanded to the size of a football field, its nucleus would be the size of a 1 mm pellet.

This is the kind of totally unexpected result that makes researchers pause, take a deep breath, and wait for someone else to stick his or her neck out to add perspective or corroboration. Although Rutherford published these results in May 1911, he did not wage an advocacy campaign to have them broadly accepted. Thomson, whose plum-pudding ox was being gored, remained skeptical and unsupportive.

If Rutherford's experiment truly established that all of the positive charge in an atom was concentrated in a tiny nucleus, then where were the electrons? Clearly, in hydrogen gas, they have to be confined somewhere within the volume of the hydrogen molecule, which would be somewhat larger than the hydrogen atom. The principal issue quickly became: Where are the electrons and how is it that the atom can exhibit its exceptional stability? After all, it is a fundamental part of the atomic hypothesis that atoms are indestructible (except for the newly discovered case of a minority of atoms that are radioactive) and maintain themselves for billions of years through many chemical and physical changes without altering their fundamental properties. It did not seem credible that the electrons were a uniformly distributed jelly within the atom, because the negative charge would be strongly attracted toward the nucleus and ultimately be pulled into be annihilated by it. What kept the electrons from being pulled into the nucleus to neutralize the positive charge?

Bohr had told Thomson that he wanted to go to Manchester to learn a little about radioactivity. Immediately after his arrival, Rutherford suggested to Bohr that he study the absorption of α-particles by aluminum, but after a few weeks' labor Bohr realized that he was much more interested in theoretical work that would rationalize the new small nucleus measurements. Both Bohr and Rutherford recognized that someone had to visualize an atomic model that incorporated a tiny dense nucleus and was more compelling than Thomson's plum-pudding model. Rutherford helped Bohr get the problem in perspective, but it would be Bohr's job alone to assemble the ideas into a coherent theory. Bohr began working in his rooming house to better concentrate on the details, and he ultimately returned to Denmark to finish his theory after a mere 3 months' stay at Manchester.

Perhaps one thing drawing him home to Denmark was the memory of a lovely smile framed by curly blond locks. Immediately after his return, he obtained an assistantship at the university and married his girlfriend of 2 years, Margrethe Nørlund. Margrethe was the daughter of a pharmacist and student at a type of Danish school specializing in home economics and nursing. Like Bohr, Margrethe was estranged from the Lutheran Church, so they were married in a civil ceremony. For a scientist like Bohr, living with one foot in the mundane world and one in the world of atoms, having a well-grounded wife would be especially important. During their long marriage, Margrethe ran the Bohr household smoothly, entertained their many visitors warmly, and joyfully raised their six sons.

Bohr's way of thinking about this quantum mechanical problem was to relate it to known classical theory, and classical physics already had a fine model of a long-lived dynamical system with an attractive and dense core: Newton's solar system. Despite the sun's gravitational force continuously pulling earth directly toward it, earth's orbit is stable because its acceleration toward the sun is exactly large enough to hold it in its nearly circular trajectory. However, applying such a planetary idea to electrons circulating about a heavy nucleus led immediately to an insurmountable problem: The well-established equations of electrodynamics required that accelerating charges, including charges in an orbit, continuously radiate energy. Therefore, such an atom, relentlessly squandering its energy, could hardly be permanent and unalterable. As its electrons emitted radiation, they could be expected to lose angular momentum and eventually spiral into the nucleus to be annihilated.

However hard Bohr and other scientists struggled, they simply could not conceive of electrons stably confined in a small atomic volume without requiring some acceleration of the electrons, again leading to the paradox of a radiating, short-lived atom.

Actually, Bohr was trying to solve an insoluble problem, because the atom cannot be described by using even the best-equipped Newtonian toolkit. Bohr was going to have to perform radical surgery on Newtonian mechanics to model the atom. Stalled after months of effort, Bohr ran across one key observation that would open his imagination to a valid quantum mechanical model of the hydrogen atom, the Balmer formula.

This formula was inspired by more than 200 years of careful observation of the light emitted from very hot atoms. Although Newton's observation that a prism could disperse white sunlight into a continuous rainbow of colors was a good first approximation, advances in the resolving power of optical instruments had shown that the sun's spectrum actually contained many intense lines. Beginning with the observations of Thomas Melvill in 1752 that burning table salt emitted a pair of brilliant yellow lines, scientists had expended vast energy in observing and cataloging spectral lines and identifying which atoms emitted them. In the laboratory, the lines could be excited by combustion of solids or liquids or electrical discharges in gases. By the mid-nineteenth century, careful measurement allowed scientists to identify microscopic quantities of atoms by comparing the spectrum of an unknown material with their catalogs [1, pp. 140–145]. Furthermore, astronomers had discovered that examination of sunlight and starlight could even reveal what materials were present in celestial bodies.

Although many scientists had speculated that these mysterious lines resulted from vibrations internal to the atoms themselves, no one had submitted a serious hypothesis that could explain *what* was vibrating. It was clear that the origins of these lines were outside the boundaries of classical physics.

Balmer had been a PhD mathematician teaching in a girls' school in Basel who inferred from just a few lines in the spectrum of hydrogen atoms a very useful general formula for the frequencies of light in these lines. His formula fit an entire series of line frequencies ν_{ab} to the expression

$$\nu_{ab} = R\left(\frac{1}{b^2} - \frac{1}{a^2}\right) \tag{9.1}$$

where $b < a$ and R was a constant that Balmer determined. The most intense lines that Balmer found were for $b = 2$ and $a = 3$, 4, 5, or 6, but Eq. (9.1) represents an infinite manifold of lines as a assumes integral values above 6. Figure 9.4 shows how the energies of these lines increase to a limit below one-quarter of a Rydberg, R, as a increases.

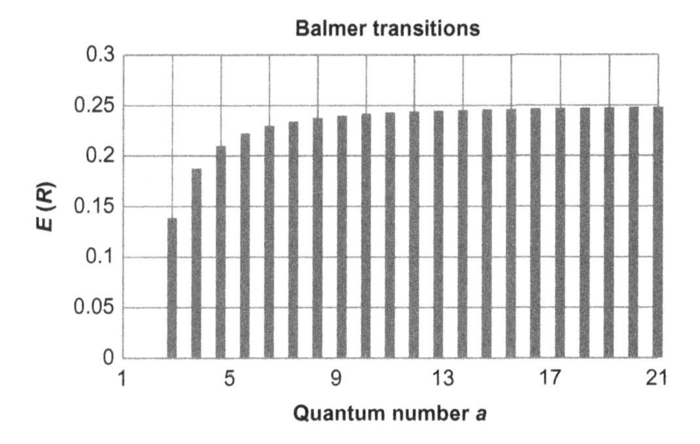

Figure 9.4 The Balmer lines. The transition energies from states of quantum number a to the $b = 2$ state are given in fractions of a Rydberg ($1R = 13.6$ eV).

Imagine how suggestive this formula was for a befuddled Bohr trying to assemble the pieces of a stupefying puzzle. Planck and Einstein had both developed ideas that maintained that the energy of a vibrating oscillator or a photon was proportional to its frequency, and the two were related through

$$E = h\nu, \tag{9.2}$$

where h was a constant that Plank determined from his radiation law (Chapter 7). So if ν_{ab} was actually proportional to an energy, then Balmer's formula suggested that the characteristic lines in the hydrogen spectrum corresponded to photons emitted when electrons moved from one energy state to another within the atom. If Bohr could only devise a scheme giving electrons bound with energies proportional to $1/a^2$, then he would be well on his way to solving the hydrogen atom mystery. Just a few short months after Bohr read Balmer's publication, he had proposed the following model for the hydrogen atom [3].

In a hydrogen atom, one electron revolves around one proton and has an infinite manifold of "stationary state" energy levels available for it to occupy when excited. Electrons making transitions from a more energetic (loosely bound) stationary state to a more tightly bound stationary state must emit light of energy described by an equation such as the Balmer formula Eq. (9.1) and Plank's relation Eq. (9.2). Each transition must emit exactly one photon. Just as in a gravitationally bound planetary system, the most tightly bound electrons circle closest to the nucleus.

To incorporate all of these pieces, Bohr made the revolutionary but necessary hypothesis that electrons circulating in the stationary states or orbits of the atom simply do not emit radiation, despite the classical prediction. The only radiation emitted is when electrons *move* from a less tightly bound to a more tightly bound stationary state.

Bohr realized that by floating such a radical hypothesis, his work would only be provisionally accepted even if it showed agreement with experiment. We will see below that it accomplished much more than he expected.

Reasoning by analogy with Plank and Einstein, Bohr assumed that in an orbital described by the *quantum number a*, the kinetic energy of the electron will be given by

$$W_a = \frac{1}{2}h\left(\frac{\nu}{2\pi r}\right)a \tag{9.3}$$

where the term in paranthesis is the angular frequency with which the electron circles the proton. This is a very significant equation because it sets up an infinitude of orbits whose energies increase as a becomes a larger and larger number, just as Planck presumed for cavity radiation (Chapter 7). The value a became known as the *quantum number* of the orbit.

You might be wondering about the factor of one-half, which is not present in the quantizations of either Planck or Einstein (Chapter 8). Bohr showed that the unexpected factor of one-half was necessary to make Eq. (9.3) consistent with the expected values of the electron's angular momentum and the total energy, but it was originally regarded with suspicion as a weakness of the theory.

To complete the Bohr theory, all that was required was to add the two equations of motion describing the force holding the electron in its orbit around the proton and the total energy of this electron [4]. To make the analogy to the Newtonian solar system clearer, Figure 9.5 contrasts the motion of a planet of mass m moving in a circular orbit at a velocity v around a sun of mass M and an electron of mass m and charge $-e$ revolving around a nucleus of charge $+e$. In describing the Newtonian orbits, G is the gravitational constant that gives the force attracting two unit masses that are a unit of length apart and M is the mass of the sun. In the Bohr orbit force calculation (9.4b), we choose the units of charge of e to be statcoulombs so that the force between unit charges separated by a unit of length is equal to one and does not show up in the equations. Equations (9.4a) and (9.4b) relate the forces on the orbital masses to their accelerations, whereas Eqs. (9.5a) and (9.5b) of Figure 9.5

Newtonian planetary orbits

$F = mA$

A gravitational force = m·(acceleration towards sun)

$$\frac{GmM}{r^2} = \frac{mv^2}{r} \tag{9.4a}$$

Total energy = kinetic + potential (9.5a)

$$E = \frac{mv^2}{2} - \frac{GMm}{r}$$

Kinetic energy in orbit (9.6a)

$$\frac{mv^2}{2} = \text{const}$$

A Bohr atomic orbits

$F = mA$

Electrostatic force = m·(acceleration towards nucleus)

$$\frac{e^2}{r^2} = \frac{mv^2}{r} \tag{9.4b}$$

Total energy = kinetic + potential (9.5b)

$$E = \frac{mv^2}{2} - \frac{e^2}{r}$$

Kinetic energy in orbit (9.6b)

$$\frac{mv^2}{2} = ah\frac{v}{4\pi r}$$

Figure 9.5 Newton's planetary equations compared to Bohr's.

compare their energies [5, pp. 39–49]. Equations (9.5a) and (9.5b) follow from Eqs. (9.4a) (9.4b) and because force is the radial derivative of potential energy (calculus alert!). These equations would be very easy for Bohr's contemporaries to accept. Equation (9.6b), which is simply a restatement of Eq. (9.3), contained the sole germ of the quantum theory and would be the only novelty.

It turns out that the algebra for solving these three equations is somewhat cumbersome. If you wish to follow it, the details are given in Figure 9.6.

The solution of Eq. (9.11) gives the energy of an electron in a stationary state of quantum number a. The rewarding feature of this equation is that the binding energy is proportional to $1/a^2$, which agrees with the Balmer formula (Equation 9.1) describing the light frequencies found in the hydrogen spectrum. Bohr had synthesized the answer he needed.

But there was to be much more. Equation (9.11) allowed Bohr to evaluate Balmer's constant R, and it agreed with Balmer's fit to the spectral lines to within experimental accuracy. Furthermore, Eq. (9.10) gave very credible numbers for the orbital radii r_a, starting around 0.0529 nm ($0.0529 \cdot 10^{-9}$ m) for the innermost orbit. To Bohr, this agreement with two key features of the hydrogen atom must have look dazzling.

Margrethe later reported Bohr's state of mind after submitting the first paper describing this work to Rutherford in March 1913. "We were eagerly awaiting [sic] after the papers went off to Rutherford—oh, we were eagerly waiting for the answer. I remember how exciting it was. And Rutherford was delighted with the first paper . . ."[1, p. 153].

Scientists, however, have rightly learned to be skeptical of theories that are built on radical assumptions. Theorists who work with math all day can learn to make numbers dance for them; who knows whether Bohr just worked out so many combinations of assumptions that he finally fabricated one lucky but rickety edifice.

However, when a theory that you suspect has been jerry-rigged together to fit the hydrogen atom fits other atoms too, then it becomes much more convincing. Bohr first pointed out that his theory could describe another puzzling set of spectral lines analyzed by Fowler in London and attributed to helium. If Bohr plugged in a charge of $2e$ instead of e for the nucleus in the above equations, then the proper frequencies to describe the helium ion spectrum came tumbling out. (Helium has two

From Eq. (9.4b)

$$v^2r^2 = e^2\frac{r}{m} \qquad (9.7)$$

From Eqs. (9.4b) and (9.5b)

$$v^2r^2 = -E\frac{2r^2}{m} \qquad (9.8)$$

From Eq. (9.3b)

$$v^2r^2 = \frac{a^2h^2}{4\pi^2m^2} \qquad (9.9)$$

Since Eq. (9.7) = Eq. (9.9)

$$r_a = \frac{a^2h^2}{4\pi^2me^2} \qquad (9.10)$$

Since Eq. (9.7) x (9.7) = Eq. (9.8) x (9.9)

$$E_a = -\frac{2\pi^2me^4}{h^2a^2} \qquad (9.11)$$

Figure 9.6 Solving Bohr's equations gives orbital radii and binding energy.

electrons, so its singly charged ion has one electron circling a doubly positive nucleus.) This series of lines, Bohr maintained, should fit a Balmer-like formula with $4 \cdot R$ as the constant instead of R.

The experimentalist Fowler had some skin in the game because, after all, he had measured the He lines. So he was the immediate authority who could certify Bohr's agreement with experiment. Fowler put on his green eyeshade and compared Bohr's He prediction with his experimental values. Bear in mind now that spectral frequencies could be measured to excellent accuracy even in the preelectronic age. Fowler returned his verdict: The agreement was not quite correct, because the constant should be $4.00163 \cdot R$ instead of $4.0000 \cdot R$. Even though the physical constants e, m, and h were not known to great accuracy, they canceled out in determining this ratio.

Bohr was probably thinking that he never expected the theory to be even as good as it was. What was all the quibbling about? In reexamining his derivations, however, he realized that he had made an approximation in setting up his dynamical equations. The helium nucleus is not infinitely heavier than the electron, only 7300 times heavier, so in the electron–helium nucleus system both particles rotate about their center of mass, making the kinetic energy just slightly larger than in the original calculation. When Bohr used this correction to calculate the Rydberg constant of the He ion, he got $4.00163 \cdot R$, in perfect agreement with Fowler's experiment.

These successes were justly regarded as a stunning triumph for Bohr. Einstein had greeted Bohr's original hydrogen paper with guarded optimism but when told of the excellent agreement with the He ion spectrum, he said with appreciation, "This is an enormous achievement. The theory of Bohr must then be right" [1, p. 154].

To summarize then: Bohr developed a model that he hoped would fit the spectral lines of the H atom. In so doing, he made several radical and, from the standpoint of classical or Newtonian physics, unjustifiable assumptions, but his theory allowed him to beautifully fit the Balmer lines of the H spectrum. As a bonus, the same theory, corrected for the finite electron mass and double charge of He, allowed him to predict key lines in the He ion spectrum to an accuracy of two parts in 1,000,000.

Stripped to its basics, the theory could even fit on a postage stamp (Figure 9.7).

Bohr began to reap the benefits of this tremendous achievement and could now turn his energies to developing his own career. The success that the Bohr atom had achieved made him impatient with his meager docentship and cramped quarters in Copenhagen. He left Copenhagen for England, accepting Rutherford's offer of a readership in Manchester, but he was always devoted to Denmark, and his real goal was to be a professor at the University of Copenhagen. Pulling as many strings as he could, he returned to a professorial appointment in Copenhagen in 1916 and began working toward establishing an Institute for Theoretical Physics almost as soon as he returned. By 1917, the Ministry of Education had approved the funds, and he busied himself with establishing this institute and making it the center of the atomic physics world.

Figure 9.7 Denmark's stamp celebrating the 50th anniversary of Bohr's theory.
Source: Post Danmark stamps, design by Viggo Bang.

Bohr initially hoped to explain the energy levels of atoms with more than one electron but was unsuccessful. In the solar system, the effects of other planets on earth are minor compared to the effect of the sun, whereas in the Bohr atom the other rotating electrons have the same repelling charge as the nucleus has an attractive charge. This "many body" problem is mathematically intractable. Bohr tried to fit extra electrons into more elaborate ad hoc stationary states but with only limited success.

So after Bohr's revolutionary advances of 1913, physics began a 13-year period of groping toward a less heuristic or ad hoc method of understanding the subatomic world than Bohr had just glimpsed. To develop the new quantum mechanics as a reliable tool, physicists needed to do some very deep thinking about the revolutionary ideas they were embracing. Bohr became the most influential physicist in the world in guiding these developments as he introduced the *correspondence principle*, which dictated that the confusing and sometimes even counterintuitive results coming from quantum mechanics should approach classical results in the limit of large particles and heavy masses. This idea allowed the intuitions developed in solving classical systems to guide understanding of quantum results. By 1921, the influential Professor Arnold Sommerfeld could describe Bohr as "director of the atomic theory" [1, p. 269]. Bohr's theory was honored when he was awarded the physics Nobel Prize in 1922.

Bohr went on to make many other contributions to physics throughout his long life. For instance, another important Bohr insight was the *liquid drop model* of the atomic nucleus, which likened a nucleus to a drop of liquid with properties such as surface tension and yielded useful insights into how stable a nucleus of a given size was and therefore how likely it was to fission.

Are you getting to feel that Bohr shunned the too abstract and was most comfortable solving problems by drawing parallels with mechanical or electrical systems that are well understood, like the solar system or the balance of forces holding a stretched liquid drop together? My junior course in atomic physics at Caltech in 1959 was taught by Tom Lauritsen, who had worked with Bohr in Copenhagen as a postdoctoral student. Evidently he had absorbed the gospel of modeling very deeply because he told us once in a philosophical moment, sucking on his ever-present pipe, "To be a good physicist, you have to not only understand how atoms work, you have to understand how an atom would work if it were made of cornmeal mush!" Lauritsen delighted us with his puckish sense of humor. He lectured one day on the shell model of the nucleus, a model (*another model!*) proposed by Maria Groppert-Mayer and colleagues (Nobel Prize, 1963) to explain why some nuclei are much more stable than others. Mayer's son Peter was a Caltech physics student sitting in our class. Lauritsen completed his description of the model and began discussing its limitations by turning to Peter, pointing his pipe at him, and continuing "Peter, here's where your mother went wrong . . .."

What physicists really wanted was the end of ad hoc quantum assumptions. They yearned for the kind of grand synthesis that Maxwell and his followers had achieved in the late nineteenth century by encompassing almost all the phenomena of electricity and magnetism in four differential equations [5]. These vector partial differential equations can be ugly to solve, but they give scientists the power and confidence to master a wide variety of problems.

Here are just a few of the ways that Bohr managed the development of quantum theory from 1913 to its final maturity in 1925.

When Bohr gave a lecture at Göttingen in 1922, a physics grad student named Werner Heisenberg raised a rather astute objection. After giving him a hesitant, provisional answer, Bohr approached the student after he concluded the lecture and asked him to walk over to Hain Mountain with him later that afternoon. During 3 hours of intense discussion, Bohr invited Heisenberg to Copenhagen. A year and a half later, over Easter 1924, the discussions continued during a 100-mile hike to Hamlet's castle in Elsinore and beyond. After these sessions, Bohr confided in a friend, "Now everything is in Heisenberg's hands." Heisenberg left Copenhagen for Göttingen in the summer of 1925 to return to his position as *privatdocent*. It was a period of ferment and mental turmoil for Heisenberg as notions of Bohr's quantum ideas swirled around in his mind, mating with possible advanced mathematical representations for particles and waves. Escaping to the pure air of the North Sea island Helgoland to combat a debilitating bout of hay fever and deeply gripped by his obsession for developing a complete quantum theory, Heisenberg began to make real progress. He derived an equation that described each problem in terms of the likelihood of each stationary state transitioning to all other quantum states [6]. By late July, the travail was complete, and Heisenberg mailed his paper on matrix quantum theory to *Zeitschrift für Physik.*

Heisenberg's formulation, relying on matrix mechanics, a branch of mathematics most physicists were not familiar with, seemed intimidatingly abstruse. Apparently, independently of Heisenberg, a Zürich professor named Erwin Schrödinger, inspired by Louis de Broglie and Einstein's concepts leading him to visualize particles as wave packets, published a *wave equation* in 1926. Using this equation, one could directly determine the effects of a given potential energy on a pure wave of constant wavelength [7]. Adding up a series of pure waves to produce a more localized wave could represent a particle. The Schrödinger equation proved to be a much more accessible foundation for the new quantum mechanics or, rather, wave mechanics.

Heisenberg, naturally, saw many deep problems with this rival formulation and pushed Bohr to invite Schrödinger to Copenhagen to compare and explore their rival methods. Schrödinger arrived by train just a few months later and was met by Bohr, who escorted him to the guest bedroom in the Bohr home. For the next few days, Bohr and Heisenberg discussed and evaluated all the aspects of the new theory in discussions with the cornered Schrödinger, beginning at breakfast and continuing until late at night. Schrödinger's health finally broke under this assault, and he returned to Göttingen to recover. He bore no ill will, however, because he wrote this description of Bohr to a friend: "honored like a demigod by the whole world, and who yet remains ... shy and diffident like a theology student."

Although painstaking mathematical investigation would later show the two forms of quantum mechanics to yield equivalent answers, Schrödinger's formulation, a comprehensible linear differential wave equation, would prove to be most tractable for most problems. The floodgates were finally opened to the use of wave mechanics in solving a broad array of quantum problems in solids and atoms more complex than hydrogen.

These theories nicely accommodated the wave–particle duality that had been required by the quantum theory, a concept Bohr defined as *complementarity* [8]. They also introduced an element of indeterminacy into physics. Heisenberg demonstrated with his *uncertainty principle* that simultaneous measurements of momentum and position of an atomic particle to perfect accuracy is absolutely impossible. Furthermore, Schrödinger's equation underscored that after two atomic particles interact, the identities of the particles became mixed up.

Legitimizing such a bizarre theory was a process that continued for many years. At the Solvay, Belgium, conference in the autumn of 1927, a gathering of notables in physics that included Einstein, Plank, Curie, and de Broglie, Paul Ehrenfest described Bohr as "towering over the everybody, at first not understood at all ... then, step by step, defeating everybody" [1, p. 316].

The debates Bohr had with Einstein (Figure 9.8) over the problems of uncertainty and indeterminacy raised and aired classic problems that physics had to overcome. In fact, physicists are still struggling with the true meaning of aspects of wave mechanics. For Einstein, the universe had to be classically deterministic; objecting to the indeterminacy built into quantum mechanics, he was quoted as saying "[God] does not play dice" [1, p. 318]. Later, in 1928, Einstein criticized wave mechanics as a "cleverly concocted philosophy."

Bohr's life was a satisfyingly productive one, but not untouched by tragedy. His dearly loved mother, Ellen, died of cancer and his sister, Jenny, suffered from mental problems, finally dying in an institution. Bohr's promising 17-year-old eldest son, Christian, was washed off the family sailing cutter by a rogue wave during a stormy outing in 1934 in full view of his father and friends. Christian was a good swimmer, and a lifebuoy was thrown in his direction, but he could not grasp it and was carried away from the boat. As Christian disappeared in the raging sea, Niels's friends had to restrain him from jumping in after the boy. Bohr had to face Margrethe with the tragic news after his boat returned to shore, and Christian's funeral had to be delayed 2 months until the body could be recovered.

Figure 9.8 Einstein and Bohr in 1930.
Source: Niels Bohr archive.

The days of World War II were painful for the Danes after the German army invaded Denmark in 1940. Bohr resisted the temptation to flee to Sweden because he was committed to safeguard the members of his institute and the security of German Jewish scientists he had aided. In defiance of the Third Reich's ban on holding gold, both Max von Laue (Nobel 1914, for demonstrating X-ray diffraction by crystals) and James Franck (Nobel 1925, for observing stationary states in mercury atoms) had left their Nobel Prize medals at the Bohr institute for safekeeping. As German soldiers marched through the streets, Bohr and his colleague George Hevesy discussed whether it was secure enough to safeguard them by burial. They finally decided to dissolve them in acid and preserve the gold solutions as innocuous reagent bottles shelved with the Institute's chemicals. The gold solutions remained undisturbed despite German occupation of the institute in 1943, and the obliging Nobel Foundation was able to recover the gold and recast the medals after the war.

As the occupation continued, it became too widely known that the expertise in Bohr's institute could play an important role in either the German or Allied atomic bomb efforts. Heisenberg had apparently gone over to the dark side to become head of the German atomic bomb program and made a well-publicized visit to his old mentor Bohr in 1941. Although Bohr and the other Danish scientists gave their old friend Heisenberg the cold shoulder, the newspapers, under occupation pressure, tried to give the impression that they were collaborating. In 1943, the British, through a microdot letter from James Chadwick delivered by an intelligence officer, extended an invitation to Bohr to come to England to help on a "particular problem." To add to the complexity of Bohr's situation, the rumor circulated that the Danish resistance was mining tunnels under Bohr's laboratory in an attempt to destroy it to deny any conceivable benefits to the Germans. Finally, when Swedish diplomatic sources notified Bohr in 1943 that the Nazis were preparing to arrest him, he fled by boat with his family to Sweden and was ultimately flown to England in a British Mosquito bomber.

Bohr contributed what he could to the Manhattan Project but did not consider his help to have been crucial. At this stage in his life, Bohr was beginning to think more like a scientific statesman than a journeyman scientist. The disciplined scientific thinker tries to illuminate current problems by visualizing how they will play out under extreme conditions and far out into the future, and Bohr was becoming worried. He had traveled to Russia many times and had good contacts there, leading him to the conviction that Stalin's paranoia would feed on the American–British atomic bomb monopoly with devastating consequences. Bohr resolved to do what he could to persuade the Allies to share some information with the Russians. It was a measure of his international reputation that he was able to arrange appointments with both Roosevelt and Churchill in 1944.

By now you have probably grasped a picture of a painfully honest man who devoted his life to an exhaustive search for the most profound truths. There is no better way to highlight Bohr's exhaustive groping for the best compromise than to contrast him with two of the greatest politicians of the twentieth century.

Bohr's friend Lord Cherwell, Churchill's scientific adviser, tried to dissuade Bohr from meeting with Churchill, as it was immediately before the D-day invasion of Europe, but Bohr was a man on a mission. When the meeting did take place, Churchill did not have the patience to listen to Bohr's broken English, and the interview went so badly that Bohr later reported, "It was terrible. He scolded us like two schoolboys" [1, p. 501]. As Churchill was shooing Cherwell and Bohr from his office, Bohr asked if he might send Churchill a memorandum, to which Churchill replied that he hoped it would not be about politics.

In the United States, things seemed hopeful at first. Bohr's friend Supreme Court Justice Felix Frankfurter was able to arrange an interview with Roosevelt about 2 months after the Churchill meeting. This time Bohr prepared a memorandum stating that the nuclear powers should forestall a "fateful competition about the formidable weapon" and "uproot any cause of distrust between the powers on whose harmonious collaboration the fate of coming generations will depend" [1]. Roosevelt trusted Stalin very much more than Churchill did and was much more receptive to Bohr's arguments during the hour-long discussion. Apparently, they had agreed to a future interview, but Churchill had other plans.

One month later, Churchill made the point to Roosevelt that he was very much worried about Bohr leaking atomic secrets to Russia. Bohr traveled too much, he talked too much, and he had too many contacts in Russia. Moreover, the worldwide respect of influential scientists gave him an independent power base outside of the

Figure 9.9 Bohr in his study, 1935.

political sphere. Churchill concluded, "It seemed to me the man ought to be confined .... I did not like the man you showed to me with his hair all over his head, at Downing Street" [1, p. 502]. Churchill's visceral antipathy ended Bohr's influence with Roosevelt.

Bohr's efforts to make a more peaceful world would continue for many years, first in negotiations with American leaders such as George C. Marshall, and then by taking his case to the public. In August 1945, he published an article in *The Times* of London calling for "free access to full scientific information and ... international supervision of all undertakings." He published a similar article in *Science*.

Bohr continued his atomic research (Figure 9.9) and efforts for world peace through the 1950s [8, pp. 288–296]. He was a strong supporter of major European scientific institutions such as CERN and Nordita. The Bohr archives show cheerful photographs of him with world notables such as Jawaharlal Nehru, Ben Gurion, King Fredrik IX of Denmark, and Louis Armstrong.

Bohr died of heart failure during his customary afternoon nap in his own bed in 1962. Few men have ever achieved and merited such worldwide respect and affection. It is a comforting thought that the same twentieth century that spawned Hitler, Stalin, and Mao could nurture this luminous spirit, this secular saint—Niels Bohr.

References

[1] A. Pais, Niels Bohr's Times in Physics, Philosophy, and Polity, Clarendon Press, Oxford, 1991, pp. 108–111.

[2] W.H. Cropper, Great Physicists: The Life and Times of Leading Physicists from Galileo to Hawking, Oxford University Press, New York, NY, 2001, p. 316.

[3] N. Bohr, On the constitution of atoms and molecules, in: A.P. French, P.J. Kennedy (Eds.), Niels Bohr: A Centenary Volume, Harvard University Press, Cambridge, MA, 1985, pp. 80–86.

[4] A.P. French, P.J. Kennedy (Eds.), Niels Bohr: A Centenary Volume, Harvard University Press, Cambridge, MA, 1985, pp. 41–45.
[5] W.D. Niven, The Scientific Papers of James Clerk Maxwell, Dover, New York, NY, 1952.
[6] K. Hannabuss, An Introduction to Quantum Theory, Clarendon Press, Oxford, 1997, p. 179.
[7] H.C. Ohanian, Modern Physics (Second Edition), Prentice Hall, Englewood Cliffs, NJ, 1995. p. 200.
[8] A.P. French, P.J. Kennedy (Eds.), Niels Bohr: A Centenary Volume, Harvard University Press, Cambridge, MA, 1985, pp. 110–112.

Bibliography

A.P. French, P.J. Kennedy (Eds.), Niels Bohr: A Centenary Volume, Harvard University Press, Cambridge, MA, 1985.
Niels Bohr, videotape by Ole John, Danish Government Film Office.
Some contemporary footage, good interviews.
P. Dam, Niels Bohr, 1885–1962, Atomic Theorist, Inspirator, Rallying Point, Royal Danish Ministry of Foreign Affairs, Copenhagen, 1985.
R. Moore, Niels Bohr, the Man and the Scientist, Hodder and Stoughton, London, 1967.
A. Pais, Niels Bohr's Times, in Physics, Philosophy, and Polity, Clarendon Press, Oxford, 1991.
Written by a man who worked with Bohr and Einstein. All the physics is here in detail.

10 Conclusions, Status of Science, and Lessons for Our Time

10.1 Conclusions from Our Biographies

By now it should be clear that I regard the scientific method not as simply the inductive process for falsifying hypotheses, but as a broad range of attitudes, criteria, and procedures that scientists bring to their tasks. The biographical chapters were intended to deepen and flesh out these abstract notions of Chapter 2 by tracing the specific thought processes of eight very gifted and creative scientists. What can we conclude by generalizing among the members of this successful group?

To begin with, these scientists were all strikingly skeptical of the conclusions of others. They all thought that it was worthwhile to swim against the current; they trusted their own observations and reasoning enough to invest their time, energy, and reputation to develop their own ideas.

But our group of scientists did not always perform the fair and judicious evaluations we proposed in Chapter 2. The story of Columbus was, of course, intended as a cautionary tale. Columbus had promised Ferdinand and Isabella that he would reach the Indies and simply could not alter that pledge. His inability to modify a hypothesis he had clearly falsified diminished his stature as the discoverer of a new world. Another obdurate scientist was Priestley, who, for all his brilliance and skill, could not put aside his personal agenda and discard phlogiston theory.

10.2 What Thought Processes Lead to Innovation?

All of these scientists deeply immersed themselves in the subject that fascinated them. It would indeed be difficult to invent new science that patches up the difficulties of the old if you were not fully versed in what the old theories were.

Note that our scientists' innovative thought processes were far from linear. Sometimes even the theorists worked backward from an answer they knew to be correct. Planck knew precisely the radiation law he needed and assembled a derivation for it that his contemporaries thought questionable. Bohr also knew how the form of his atomic energy levels must depend on the quantum numbers defined by the Balmer spectral formula, giving him a vital clue in deriving his model.

How the Great Scientists Reasoned. DOI: http://dx.doi.org/10.1016/B978-0-12-398498-2.00010-2

All of these men were committed to their profession. Columbus devoted his life to his quest; from his youth to his death, he ceaselessly prepared for and campaigned for his route to the Indies.

Two of our scientists, Priestley and Faraday, became master craftsmen of experimental science. It is a particular pleasure to read their laboratory notes and follow their thought processes as they observed, formulated, tested, and revised their ever-expanding thought processes. Their approaches were open and sometimes even playful; their observations gradually expanded as their experiments dictated. Lavoisier was no less a master craftsman, but his temperament made him more of an architect and builder, and his *Elements of Chemistry* is closer to a series of mathematical proofs.

Our theorists were expected to build logically on previous work, but Einstein may be an exception to this principle because it has been hard for historians to evaluate his exact debts to his predecessors. He seems to have suppressed the extent of his knowledge of theory and experiment in a number or areas: Brownian motion, the Michelson–Morley experiment, and the theoretical work of Poincaré.

10.3 Is the Scientist an Outsider?

One partial thread running through these biographies was that many of these scientists were outsiders. Columbus was a wanderer who very quickly shuffled off, and sometimes even attempted to conceal, his Genoese roots. Priestley was a religious dissenter barred from the mainstream of English life. Faraday was another dissenter and an ambitious and energetic poor boy determined to leave his class behind. Röntgen's quest for education forced him from his native Holland to Switzerland, while later professional opportunities led him to Germany. Einstein was temperamentally the most confirmed outsider. As a Jew and a pacifist, he was eager to escape the military demands of his German citizenship. He was comfortable moving to Switzerland, Czechoslovakia, Germany, and America. Even as a scientist he positioned himself as a dissenter from the prevailing views on quantum mechanics. Many of the papers he published were intended not to confirm but to revolutionize science.

A clear exception was the insider Lavoisier, who might even be classified as a pillar of the French establishment. As far as chemistry was concerned, however, he was an energetic and committed reformer, an intellectual bomb thrower.

Max Planck was a thoroughgoing insider by temperament. Nothing about Planck seemed to indicate that he was either an outsider or a revolutionary. By his own admission, he wanted to do "normal science," not "revolutionary science," and he was clearly uncomfortable when his discoveries led him to spark the quantum revolution. In fact, he even resisted the expansion of his brilliant innovation, quantization, to light and later to electron energy levels in atoms. Most scientists would be delighted to see their discoveries applied in new areas.

Niels Bohr was another scientist very comfortable with his upper-middle-class role and connections; his commitment to Denmark was central to his life. It is hard to see any of the attitudes of an outsider in him, except his enthusiasm for scientific innovation.

It would be difficult to come to a useful conclusion from this limited selection of men. Perhaps the eye-catching number of outsiders in our biographies may simply be consistent with high achievers in other, nonscientific fields.

10.4 The Status of the Modern Scientific Enterprise

The current state of science is healthy. Those whose lives we have celebrated and many other men and women have bequeathed us a functioning system for developing, testing, verifying, and diffusing scientific knowledge and a solid core of information on which to build. Science has progressed in the period that we have surveyed from a mostly European enterprise engaged in by a few academics and hobbyists to one that spans the globe and engages millions.

Science, however, is a profoundly human enterprise, and it will always struggle against human failings. For example, the development of new drugs is so expensive (and lucrative) that the temptation to fudge the clinical trial data verifying the benefit of these drugs is strong. It is difficult for even the most skilled reviewer to assess a mass of statistics, and the scientist who is tempted to cheat knows that it will be a long time before his skullduggery will be detected. Such violations of ethics are regularly unearthed, and there does not seem to be a way to stop them. It seems inherent that conclusions that must be certified by large-scale clinical trials will always be more susceptible to fraud than those based on understandable theory or laboratory experiments of manageable difficulty.

As physics matures, it has begun exploring some new and controversial territory. We have described the failure of Faraday in the 1850s to experimentally show that the forces of gravitation and electromagnetism have similar roots. We also mentioned that Einstein attempted at the twilight of his career to devise a theory to incorporate these two forces into one *unified field theory*. Modern efforts to concoct a theory compatible with these two forces plus the two forces present in the atomic nucleus have led to the spectacularly complex *string theory*, which treats the fundamental particles as vibrations of very tiny, very tightly stretched strings [1]. The complexity of this theory and the number of pathways it opens to exploration are beguiling. However, tests against experiment in the lower energy ranges currently accessible have not been convincing. Physics is thus on the verge of developing an important branch less tightly coupled to experiment than is traditional. Since the successes of physics in the last four centuries have been due to lusty, full body contact with observation and experiment, we must vigilantly maintain this tradition.

Another area of controversy is computer simulations. These programs can become very complex and are not solely derived from first principles. They can incorporate a large set of assumptions, any one of which might affect the ultimate answer to an unknown degree. Properly done, such simulations can be a useful tool, but the effects of each input must be thoroughly explored and vetted. Climate simulations have been recently criticized for their assumptions about how human-generated CO_2 will increase cloud cover to cause global warming [2].

10.5 Lessons for Our Time

The scientific method as described and exemplified here is the best tool humankind has developed for extracting durable knowledge, i.e., knowledge we can depend and build on. Although brilliant men such as Euclid and Pythagoras were able to erect impressive edifices of deductive and coherent mathematical thought thousands of years ago, using the full force of the scientific method as a laboratory tool to unravel the complexity of nature's physical, chemical, and biological laws is just a few hundreds of years old.

To reduce the scientific method to its bare bones, useful guesses about a topic are teased to yield specific predictions. These predictions are compared with observation. If the two do not coincide, then the original guesses must be modified.

This formulation emphasizes that the scientific method is built around prediction. A hypothesis that cannot make predictions is not falsifiable and cannot be tested. Therefore, it cannot become a link in the chain of scientific reasoning. However, any prediction can ultimately be evaluated by comparing with observation. Developing hypotheses suitable for making predictions one can test is the best recipe for reliable thinking.

Another point of surpassing importance is implicit in this reasoning: The scientific method requires meticulous examination of experimental facts or observations and careful evaluation of hypotheses. These exacting thought processes can be tedious but are vital.

10.6 Can the Scientific Method Be Applied to Public Policy?

Most of our political discourse is deliberately couched in poorly defined terms that resist examination. What is, for instance, "fair"? However, politicians are occasionally rash enough to make concrete predictions based on their policies, and these predictions thus become testable hypotheses. Each prediction can tell us how attuned to reality the public figure is. Politicians can justly expect that their voters will be forgiving of inaccuracies in complex arenas such as economics and warfare. In these areas, there will always be extenuating circumstances for which they may beg dispensation. However, it should be up to the voters to decide when a leader's hypotheses are so egregiously in conflict with reality that the politician should lose credibility. Voters searching for someone with an insightful working grasp in the area of discrepancy will be justified in looking elsewhere.

Occasionally, religious figures are rash enough to advance a falsifiable hypothesis. Leaders of major religions are usually too prudent to hazard such guesses, but smaller and more zealous groups occasionally enliven our public discourse by hazarding a specific date for the end of the world. In case you have not checked recently . . . it's not over yet!

10.7 Why So Little Interest in Science?

Here is a troubling problem: In an age when we have achieved through our knowledge of natural science unparalleled control over our world, why are so many uninterested in understanding scientific principles? Three major problems might account for some of this avoidance:

- First, ours is an age with one foot still in the romantic era. We are still in reaction against discipline, and the practice of science demands discipline.
- Second, science is intellectually demanding. The proportion of young people in the United States choosing to study physics has been dropping for at least 10 years. It is now acceptable to excuse oneself from the arduous duty of attaining literacy in science by rationalizing that, "I have no talent for it."
- Third, salary levels of scientists are modest relative to the intellectual gifts and effort required to master a science. By contrast, in many parts of the developing world, a degree in science is a ticket to immigration and rewarding employment with a first-world job. Therefore, young Americans in universities find themselves in competition with a large pool of highly motivated foreign students.

Our intellectual and journalistic classes seem to share both a profound ignorance and a general lack of interest in science. To them, science is performed by nerdy geeks who communicate poorly and spend too much time alone working on abstruse equations or bubbling flasks. Only the occasional stars like Stephen Hawking or Jane Goodall who can fire our imaginations are widely celebrated. Judging from the articles in the daily papers, ignorance of basic scientific principles such as conservation of energy and mass or the importance of selecting the proper units of measurement is common among journalists.

10.8 Knowledge Is Never Complete

Finally, our meditations on the scientific method must close with the caution that no scientific theory can be complete or proven to be true in all cases. The only exceptions would be self-contained and simple systems such as Euclidean geometry that can be built on a few solid axioms. Theories that describe the complexity of Nature must remain open to accommodate answers to questions we have not thought to ask yet.

Many people will remain uncomfortable with the uncertainty that the scientific method brings. Open-ended and provisional truth can be unnerving, but it is the best truth we have. Perhaps, after all, it is the surest way for us to maintain the appropriate reverence for the richly elaborated universe we inhabit.

References

[1] B. Greene, The Elegant Universe: Superstrings, Hidden Dimensions, and the Search for the Ultimate Theory, W.W. Norton, New York, NY, 1999.

[2] S.G. Philander, Is the Temperature Rising? The Uncertain Science of Global Warming, Princeton University Press, Princeton, NJ, 1998, p. 191.

Printed in the United States
By Bookmasters